AF389152

Recent Advances in Energy Efficiency of Buildings

Recent Advances in Energy Efficiency of Buildings

Editor

Montserrat Zamorano

MDPI • Basel • Beijing • Wuhan • Barcelona • Belgrade • Manchester • Tokyo • Cluj • Tianjin

Editor
Montserrat Zamorano
University of Granada
Spain

Editorial Office
MDPI
St. Alban-Anlage 66
4052 Basel, Switzerland

This is a reprint of articles from the Special Issue published online in the open access journal *Applied Sciences* (ISSN 2076-3417) (available at: https://www.mdpi.com/journal/applsci/special_issues/Energy_Efficiency_Buildings).

For citation purposes, cite each article independently as indicated on the article page online and as indicated below:

LastName, A.A.; LastName, B.B.; LastName, C.C. Article Title. *Journal Name* **Year**, *Volume Number*, Page Range.

ISBN 978-3-0365-4909-5 (Hbk)
ISBN 978-3-0365-4910-1 (PDF)

Cover image courtesy of Montserrat Zamorano

Contents

About the Editor

Montserrat Zamorano

Dr. in Civil Engineering at the University of Granada, with extensive teaching and research experience in environmental technology science, including topics such as environmental engineering, recycling construction and demolition waste, and energy efficiency in buildings. He has supervised a total of 15 PhD theses and published 101 scientific documents in international journals, with a total of 2588 citations by 2261 documents, 1601 of them from the last 5 years (2018–2022). These data imply an h-index of 29. He is a founding partner of the spinoff PROMA Proyectos de Ingeniería Ambiental S.L., a company that specializes in technical assistance, advice, research and development and the application of results in the field of engineering of urban services and smart cities. This professional career was recognized on February 28, 2020, when theAndalusian Medal for Environmental Merit was awarded by the Ministry of the Presidency, Public Administration and the Interior of Andalusia Government.

Editorial

Special Issue: Recent Advances in Energy Efficiency of Buildings

Montserrat Zamorano

School of Civil Engineering, University of Granada, 18071 Granada, Spain; zamorano@go.ugr.es

Citation: Zamorano, M. Special Issue: Recent Advances in Energy Efficiency of Buildings. *Appl. Sci.* **2022**, *12*, 6669. https://doi.org/10.3390/app12136669

Received: 24 June 2022
Accepted: 29 June 2022
Published: 1 July 2022

Publisher's Note: MDPI stays neutral with regard to jurisdictional claims in published maps and institutional affiliations.

1. Introduction

Buildings are important consumers of energy; in fact, they represent 30–45% of global energy use and one-third of total greenhouse gas (GHG) emissions [1,2] and make an important contribution to the urban heat island effect (UHI). In this context, the correct design and execution of buildings, the adequate use of energy sources, and the development of new materials are necessary to reduce energy demand and operating costs, as well as to reduce the emission of greenhouse gases during the life cycle of buildings.

This special issue collects a set of papers which show important efforts focused to improve the energy-efficient of buildings, in accordance with circular economy policies and their following life cycle phases: (i) the design phase, taking into account climatic parameters and climatic zones; (ii) the construction phase, using new materials for building, which include recycled waste; and (iii) the use phase, reducing energy consumption. These initiatives are summarized below.

2. Design Phase of Building: Analysis of Climate-Oriented Research in Building

Many aspects of the construction and operation of buildings depend on climatic parameters and climatic zones, so they are fundamental for adapting to and mitigating the effects of climate change [3]. In fact, a broad spectrum of themes directly and indirectly related to climate and climatic zones and buildings have shown that 88% of all climate-oriented investigations are within the scope of the general topic of energy conservation [3]. Consequently, a thorough understanding of all climate-dependent aspects will help in designing dwellings appropriately in different climate zones.

In this sense, and in the current context of the climate crisis, it is essential to design buildings that can cope with climate dynamics throughout their life cycle because it will ensure the development of sustainable and resilient building stock. Because of the higher global and surface temperature during the last 100 years, and projections of future climate conditions, it is necessary to determine the setting of new energy budgets and to transform the energy performance of buildings and cities [4,5]. To do that, and given the importance of precision in the assignment of a climatic zone when correctly sizing domestic hot water, heating and cooling systems, and the appropriate selection of the construction materials used, it is essential to take climate dynamism into account in the building design phase [6].

Thus, in the case of the cities of peninsular Spain, the allocation of climatic zones currently included in the Edification Technical Code [7] is not suitable for current and future climatic conditions. In fact, taking into account the climate data recorded in the 2015–2018 period, 80% of cities today have a different climate zone to that of the CTE [6]; moreover, it is expected that by the year 2085 and under the forecasts recorded in the RCP 4.5 and RCP 8.5 scenarios, practically all cities in mainland Spain observe a change in their climate zone to a warmer one [6]. It should be noted that the consequences observed in peninsular Spain can be extrapolated to other areas, so it is possible to conclude that significant climate change will reduce the heating energy demand of dwellings, but increase the demand for cooling. Therefore, architectural and construction standards must adapt to the urban environment's actual conditions and consider the main scenarios, in order to lead to a building design that mitigates climate change and adapts to it. Hence, it is necessary to develop new climate zones and build recommendations to preserve the correct thermal conditions of future periods [6].

3. Construction Phase of Building: Using New Materials for Building That Include Recycled Waste

The creation of economic and environmentally friendly materials based on the integration of secondary materials as substitutes represents an opportunity to reduce waste disposal and the consumption of natural resources in the building industry.

For example, innovative porous materials (PM) could be synthesized by bottom ash and silica fume (as raw materials) to promote a circular economy, and new solutions could be proposed to improve urban air quality, reducing PM concentrations [8]. Another example is the integration of glass in the fabrication of ceramic materials as fine particles with optimal dosage and particle size, in order to enhance their physical, mechanical, and thermal properties and improve the solar reflectance performance [9]. The design of ceramic tiles with recycled glass could contribute to the production of friendly construction materials to be applied in the design of cool surfaces [10], i.e., surfaces with reflective materials and coatings that reflect the solar energy radiation that hits building envelopes and urban areas [11], including roofs, facades, and pavements. Such materials are able to reduce the thermal infrared radiation outflow in the atmosphere, as well as the temperature and the solar heat gain, in order to palliate the UHI effect [12].

4. Use Phase of Building: Reducing Energy Consumption

Finally, the household sector plays a significant role in energy conservation and environmental sustainability in terms of using energy-saving goods [13], so products that swiftly dissipate energy are considered important in achieving efficiency and reducing carbon emissions [14]. However, energy saving in the residential sector is dependent on users' technological and habitual behavior; in fact, energy-efficient home appliances (EEHA) offer more energy efficiency and sustainability than the habitual correction of turning off appliances when not in use [15]. In this sense, a significant relationship between environmental concern, environmental knowledge, subjective norms, eco-labeling, and attitude towards buying has been detected. It has been also confirmed that the green self-identity moderates the existent relationship between the attitude and buying intention of energy-efficient home appliances, while environmental knowledge does not [16].

Funding: This research received no external funding.

Conflicts of Interest: The author declares no conflict of interest.

References

1. Marique, A.-F.; Rossi, B. Cradle-to-grave life-cycle assessment within the built environment: Comparison between the refurbishment and the complete reconstruction of an office building in Belgium. *J. Environ. Manag.* **2018**, *224*, 396–405. [CrossRef] [PubMed]
2. Pal, S.K.; Takano, A.; Alanne, K.; Siren, K. A life cycle approach to optimizing carbon footprint and costs of a residential building. *Build. Environ.* **2017**, *123*, 146–162. [CrossRef]
3. Verichev, K.; Zamorano, M.; Salzar-Concha, C.; Carpio, M. Analysis of Climate-Oriented Researches in Building. *Appl. Sci.* **2021**, *11*, 3251. [CrossRef]
4. Troup, L.; Eckelman, M.J.; Fannon, D. Simulating Future Energy Consumption in Office Buildings Using an Ensemble of Morphed Climate Data. *Appl. Energy* **2019**, *255*, 113821. [CrossRef]
5. Bellia, L.; Mazzei, P.; Palombo, A. Weather Data for Building Energy Cost-Benefit Analysis. *Int. J. Energy Res.* **1998**, *22*, 1205–1215. [CrossRef]
6. Díaz-López, C.; Jódar, J.; Verichev, K.; Rodriguez, M.L.; Carpio, M.; Zamorano, M. Dynamics of Changes in Climate Zones and Building Energy Demand. A Case Study in Spain. *Appl. Sci.* **2021**, *11*, 4261. [CrossRef]
7. Ministry of Housing. *Royal Decree 314/2006 of March 17, Approving the Technical Building Code (Real Decreto 314/2006, de 17 de Marzo, Por El Que Se Aprueba El Código Técnico de La Edificación)*; Ministry of Housing: Madrid, Spain, 2006.
8. Cornelio, A.; Zanoletti, A.; Braga, R.; Depero, L.E.; Bontempi, E. The Reuse of Industrial By-Products for the Synthesis of Innovative Porous Materials, with the Aim to Improve Urban Air Quality. *Appl. Sci.* **2021**, *11*, 6798. [CrossRef]

9. Mourou, C.; Zamorano, M.; Rúiz-Padillo, D.P.; Martín-Morales, M. Cool Surface Strategies with an Emphasis on the Materials Dimension: A Review. *Appl. Sci.* **2022**, *12*, 1893. [CrossRef]
10. Mourou, C.; Martín-Morales, M.; Zamorano, M.; Rúiz-Padillo, D.P. Light Reflectance Characterization of Waste Glass Coating for Tiles. *Appl. Sci.* **2022**, *12*, 1537. [CrossRef]
11. Pisello, A.L. State of the art on the development of cool coatings for buildings and cities. *Sol. Energy* **2017**, *144*, 660–680. [CrossRef]
12. Gao, Y.; Xu, J.; Yang, S.; Tang, X.; Zhou, Q.; Ge, J.; Xu, T.; Levinson, R. Cool roofs in China: Policy review, building simulations, and proof-of-concept experiments. *Energy Policy* **2014**, *74*, 190–214. [CrossRef]
13. Ali, S.; Ullah, H.; Akbar, M.; Akhtar, W.; Zahid, H. Determinants of consumer intentions to purchase energy-saving household products in Pakistan. *Sustainability* **2019**, *11*, 1462. [CrossRef]
14. Tan, C.-S.; Ooi, H.-Y.; Goh, Y.-N. A moral extension of the theory of planned behavior to predict consumers' purchase intention for energy-efficient household appliances in Malaysia. *Energy Policy* **2017**, *107*, 459–471. [CrossRef]
15. Frederiks, E.R.; Stenner, K.; Hobman, E.V.; Fischle, M. Evaluating energy behavior change programs using randomized controlled trials: Best practice guidelines for policymakers. *Energy Res. Soc. Sci.* **2016**, *22*, 147–164. [CrossRef]
16. Li, Y.; Siddik, A.B.; Masukujjaman, M.; Wei, X. Bridging Green Gaps: The Buying Intention of Energy Efficient Home Appliances and Moderation of Green SelfIdentity. *Appl. Sci.* **2021**, *11*, 9878. [CrossRef]

Article

Analysis of Climate-Oriented Researches in Building

Konstantin Verichev [1,2,3], Montserrat Zamorano [1,4], Cristian Salazar-Concha [5] and Manuel Carpio [3,6,*]

[1] Department of Civil Engineering, University of Granada, Severo Ochoa S/N, 18001 Granada, Spain; konstantin.verichev@uach.cl (K.V.); zamorano@ugr.es (M.Z.)

[2] Institute of Civil Engineering, Universidad Austral de Chile, General Lagos 2050, 5111187 Valdivia, Chile

[3] Department of Construction Engineering and Management, School of Engineering, Pontificia Universidad Católica de Chile, Avenida Vicuña Mackenna 4860, Santiago 7820436, Chile

[4] PROMA, Proyectos de Ingeniería Ambiental, S.L., 18010 Granada, Spain

[5] Administration Institute, Universidad Austral de Chile, Independencia 631, Valdivia 5110566, Chile; cristiansalazar@uach.cl

[6] UC Energy Research Center, Pontificia Universidad Católica de Chile, Avenida Vicuña Mackenna 4860, Santiago 7820436, Chile

*   Correspondence: manuel.carpio@ing.puc.cl

Abstract: Many factors and aspects of the construction and operation of buildings depend on climatic parameters and climatic zones, so these will be fundamental for adapting and mitigating the effects of climate change. For this reason, the number of climate-oriented publications in building is increasing. This research presents an analysis on the most-cited climate-oriented studies in building in the period 1979–2019. The main themes, the typologies of these investigations and the principal types of climatic zoning used in these studies were analysed through bibliographic and manual analysis. A broad spectrum of themes directly and indirectly related to climate and climatic zones and buildings was demonstrated. It was found that 88% of all climate-oriented investigations, to one degree or another, are within the scope of the general topic of energy conservation. A thorough understanding of all climate-dependent aspects will help in designing dwellings appropriately in different climate zones. In addition, a methodology that facilitates the establishment of a typology of climate-oriented research is presented. This typology can be used in future research in different scientific areas. It was also revealed that the climate zones of the National Building Codes of China, the USA and Turkey prevailed in the studies analysed.

Keywords: climate-oriented; buildings; building construction; climate zones; climate change; bibliometric

Citation: Verichev, K.; Zamorano, M.; Salazar-Concha, C.; Carpio, M. Analysis of Climate-Oriented Researches in Building. *Appl. Sci.* **2021**, *11*, 3251. https://doi.org/10.3390/app11073251

Academic Editors: Joan Formosa Mitjans and Luisa F. Cabeza

Received: 10 February 2021
Accepted: 2 April 2021
Published: 5 April 2021

Publisher's Note: MDPI stays neutral with regard to jurisdictional claims in published maps and institutional affiliations.

1. Introduction

The construction industry, in general, and the building industry, in particular, are characterised by a high level of consumption of resources and emissions, which greatly impacts the environment [1,2]; 40% of the total energy consumption and one-third of the total greenhouse gas (GHG) emissions are associated with the construction and operation of buildings [3–5]. These GHGs, mostly CO_2, are emitted throughout the entire life cycle of buildings, although typically, the operation phase accounts for the majority of energy consumption and emissions compared to the construction and demolition phases [6–8]. It can be said that, currently, the construction industry is responsible for at least one-third of the anthropogenic effect of planetary climate change. Furthermore, global GHG emissions associated with building construction are expected to double by 2050 [9].

Given this problem, numerous investigations have been developed on this subject. Thus, scientific production in the field of buildings and climate change has been increasing in recent years, in addition to becoming more interdisciplinary. Among the published aspects is the development of methodologies that allow the use of current meteorological data for the energy simulation of the dwellings [10,11], as well as the development of works that are aimed at improving the hierarchy of climatic zones in buildings [12,13], so as to reduce the construction of inefficient houses from an energy point of view, and that take into account the urban, mesoclimatic and microclimatic conditions for buildings [14–18].

In the context of the evolution of knowledge about the climate system, as well as a knowledge evolution about the anthropogenic impact on the climate, research on minimising energy consumption in building construction by optimising the use of materials and GHG emissions has begun [19–21].

The development of climate change projections has sparked interest in conducting research on the change of climate zones for buildings in future climate conditions [22,23]. Furthermore, papers have been published on the analysis of changes in the total energy consumption of housing, as well as on the possible evolutions of changes in energy consumption for the cooling and/or heating of dwellings in different parts of the world [24–28]. In addition, other research has dealt with changes in the climate values of heating degree-days (HDD) and cooling degree-days (CDD) in the future [29,30]. Likewise, studies have been carried out on the analysis of adaptation techniques [31,32] and mitigation [33,34] in relation to climate change. The concern about climate change by researchers, public administration and society in general has promoted new research on the development of state standards and programs to control the negative effects of the building construction industry on the environment and policies to mitigate the effects of climate change [35–37].

Generally, all studies in building carried out based on climatological parameters, or in the context of climate change effects, can be identified as climate-oriented. Furthermore, for the most part, studies in building are characterised by pronounced localism, as it is generally implemented on a country, region or city scale. To a large extent, this situation is due to the research focus on the final consumer (government of countries, regions, municipal authorities, local construction companies, etc.). This is a substantial factor, as local building experts have a clear understanding of the particular characteristics of the region in which they work, conduct research and carry out applied projects. However, this may be a limiting factor if we consider the development of certain unified global techniques and methodologies. The climate factor clearly emerges from localism, which in turn affects the development of climate-oriented scientific works in building.

The studies mentioned above are directly linked to climate, with topics on climate zones for buildings, meteorological data for energy simulations and change in energy consumption under conditions of global climate change. However, there are also works that are indirectly related to climate, that is, studies that do not directly address climate research, but whose results or methodologies depend on climate or climate zones. For example, research on the composition and structure of the substrate of green roofs, where the authors show that the dependence on the optimal composition and structure of the substrate for better bioproductivity of the plants, is ultimately affected by the climatic zones [38], the selection of the best hybrid cooling system to minimise energy consumption and building emissions shows how each climate zone corresponds to a technically different optimal system [39], studies on passive building cooling have shown the importance of the climate zone with respect to the various techniques used [40] and the neutral temperature of the adaptive thermal comfort model and its relation to the minimum mortality temperature have a clear regional dependence [41].

For the above reasons, the dependence of the objectives, methodologies and/or results of studies on climate is based on the dependence on climate parameters and their variability, both temporal and geographical. In addition, a certain set of climatic parameters (at certain intervals) characterises some climatic zones. Therefore, the term climate-oriented is directly related to climate zones. Thus, in the works mentioned above, it can be seen that many aspects of building construction depend on climatic zones, specialised building zones or bioclimatic zones. Therefore, an understanding of all topics, parameters, techniques and methods of building according to climate zone will help in optimally designing buildings in different parts of the world.

Due to the large number of publications dealing with construction from a climate or climate-oriented point of view, either directly or indirectly, the main objective of this work is to establish a scientific map of climate-oriented studies in building, through a bibliographic and manual analysis of the most-cited publications in the period 1979–2019, and to identify the main topics, the typologies of these investigations and the main types of climatic zoning used in these studies and the principal climate zones in which the investigated works were implemented.

2. Materials and Methods

2.1. Bibliometric and Bibliographic Methods

In recent years, different methods of bibliometrics have been widely used to analyse the development of different fields of science. Bibliometrics is divided into two main areas:

- Performance analysis, where several scientific producers are evaluated using bibliographic data and applying bibliometric index (h-index, hg-index, etc.).
- Analysis of science maps, where the structural and dynamic aspects of the field of science and its temporal evolution are studied, while also analysing the intellectual connections and their evolution in the field of knowledge [42].

To study a scientific map, it is necessary to create and analyse bibliographic networks. In the study of Moral-Munoz et al. [42], a description of the main types of bibliographic networks is presented, including the following:

- Co-citation networks—two publications can be considered as co-cited if a third publication quotes both. The strength of the co-citation relationship will depend on the number of publications citing both publications together (Figure 1). Papers A and B are associated because they are co-cited in a reference list of papers C–E. Therefore, the greater the number of co-citations, the stronger the co-citation relationship [42,43].
- Bibliographic coupling networks—in this case, two publications are bibliographically coupled if both publications cite a third (Figure 1). Papers A and B are bibliographically coupled because they have common cited paper C–E in their reference list. A higher number of references shared by two publications indicates a stronger bibliographic coupling between them [42,44,45].

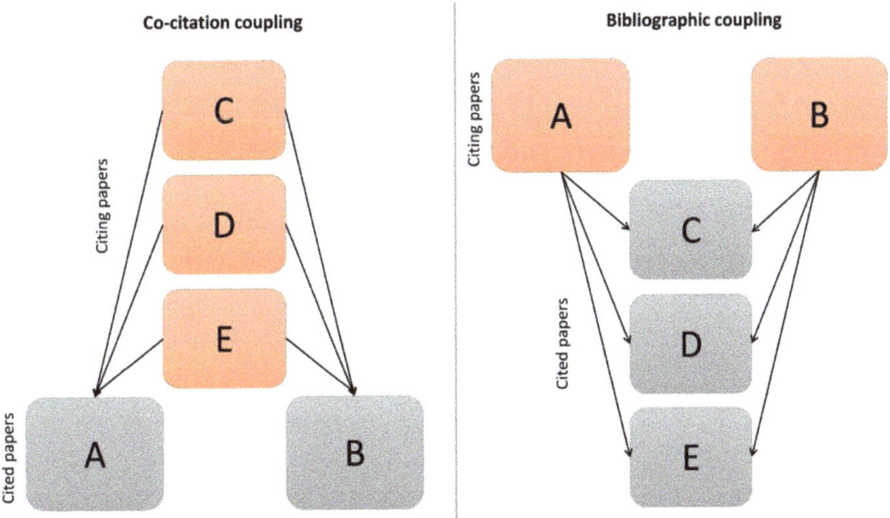

Figure 1. Bibliographic and co-citation coupling difference.

Small and Koenig, in 1977, clustered journals based on the bibliographic coupling method, demonstrating good results in comparison with the manual classification method [45]. In addition, the claim that the bibliographic coupling method is capable of grouping documents of similar research types has been confirmed in Jarneving's research [46]. Furthermore, in the work of Boyack and Klavans [46], the use of bibliographic coupling showed better results in the clustering process, compared to co-citation analysis and direct citation analysis. For these reasons, in this paper, it was decided to use this methodology for science mapping analysis.

2.2. Data Collection

This work is based on the main articles on climate-oriented studies in building. Papers published on the Web of Science (WOS) up to the end of 2019 were analysed. To find the articles, the following formula was used in "advanced search":

TS = ("climate zone" and "building") OR TS = ("climatic zone" and "building") OR TS = ("climate zones" and "building") OR TS = ("climatic zones" and "building") OR TS = ("climate zone" and "buildings") OR TS = ("climatic zone" and "buildings") OR TS = ("climate zones" and "buildings") OR TS = ("climatic zones" and "buildings") OR TS = ("climatic zoning" and "building") OR TS = ("climatic zoning" and "buildings").

The search included all languages and all types of documents for all years of operation of the WOS (until the end of 2019). Citation Indexes: SCI-EXPANDED, SSCI, A&HCI, CPCI-S, CPCI-SSH, BKCI-S, BKCI-SSH and ESCI.

The initial sample drawn from WOS included a total of 1403 documents. First, manually, the abstracts and keywords of all publications were reviewed, and publications not related to the building scientific field were excluded. After this additional manual examination of publications, a total of 1280 documents remained. These 1280 documents covered the publication period from 1979 to 2019.

To select the publications with the most significant impact, 10% of the most cited papers were used (with more than 35 citations in WOS at the end of 2019). After applying this filter, 128 articles were selected covering the period from 1995 to 2018.

2.3. Clustering Tools and Methodology

In the first stage of data processing, the Bib Excel tool was used to transform the .txt file of the WOS to a Citation Network File (with a number of commonly cited papers in references for each pair of articles). Of the 128 articles that made up the database, only 114 presented common references and therefore formed the basis of our cluster analysis.

To analyse the Citation Network File, it was processed in the program Gephi, version 0.9.2 [47]. Gephi is open-source software for graphics and network analysis, which provides the possibility of visualisation and research of all types of complex networks and systems, dynamic and hierarchical graphics. Gephi was implemented for a bibliometric analysis of research in different fields of science, applying different methods for analysis [48–50]. Note that Gephi is the most reliable software for scientific map analysis [42]. In this research, a 114×114 matrix was formed in Gephi with common references numbers for each pair of papers.

For clustering of the article network, a modularity cluster detection algorithm was used. The cluster definition algorithm is heuristic, based on modularisation optimisation and was first published in 2008 by Blondel et al. [51]. This algorithm (or Louvain's method [51]) has several advantages, such as the ease and speed of implementation and the ability to analyse large and weighted networks. In comparison with other methods, Blondel's algorithm showed better results for the definition of clusters [51,52]. The general methodology of this research is presented in Figure 2.

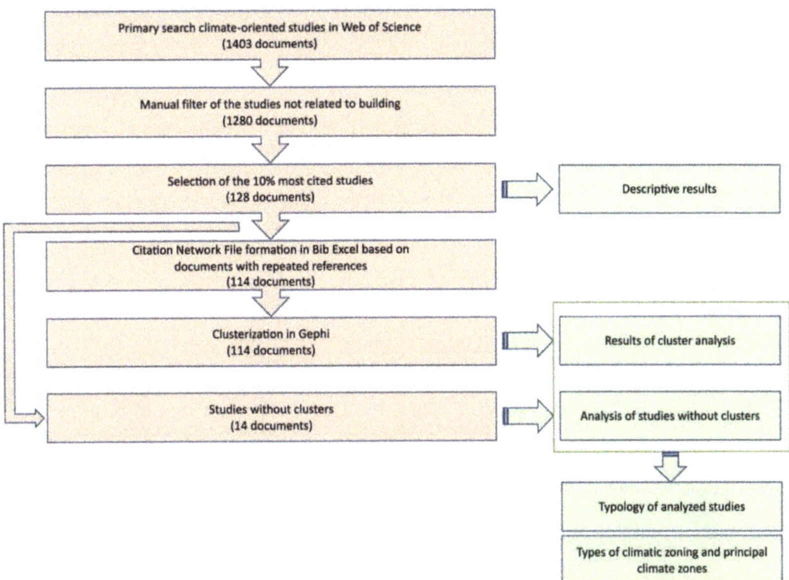

Figure 2. The general methodology of the present research.

The scope of the present research to identify the scientific map of climate-oriented publications can be summarised as follows:

- Only WOS publications were used.
- Ten percent of the most frequently cited papers were analysed.
- A bibliographic coupling method was used for clustering.
- Works published from 1995 to 2018 were analysed.

3. Results and Discussion

3.1. Descriptive Results

The time trend of the 128 publications used in this paper, all with more than 35 citations, is presented in Figure 3. It shows an upward trend, with a maximum number of documents published in 2012. The first work was published in 1995, and in the following 15 years (period: 1995–2011), 60 papers were published, while in a more recent period of only seven years (2012–2018), a similar number has been published, specifically 68 articles. The typology of the works studied includes 112 full research articles, 10 reviews, four article proceeding papers, one review book chapter and one editorial; these results once again highlight the intensive research being carried out on the subject under study in the present work.

Note that the earliest studies were related to the analysis of the effect of climate change on variations in the energy consumption of buildings [53], as well as to the analysis of the method of energy conservation by improving the thermal envelope of buildings [54]. The maximum of mean number of citations in Figure 3 in 2002 is associated with the publication of one of the fundamental articles on adaptive thermal comfort [55]. Noted that in the period up to 2005, principal studies related to the study of microorganisms in the indoor air were conducted [56,57]. Furthermore, in 2005, some of the main works on the study of cool roofs as energy saving techniques were published [58,59]. In 2008, a study on Fanger's thermal comfort analysis was published [60]. In 2014, work on the pairing of indoor thermal comfort and an analysis of the energy saving potential was published [61]. The most recent work analyses the reduction of carbon emissions from office buildings as a result of the effective building energy efficiency policy in China [62]. In general, over the past decades, the dominant theme of most of the papers under study has been energy saving in buildings. At the same time, this topic has gradually absorbed new emerging topics, for example, the topic of adaptive thermal comfort has been integrated into the energy saving concept. Urban heat island mitigation techniques, systems optimisation, building envelope optimisation techniques and emission reduction techniques have also become progressively more complex. Meanwhile, the analysis of microorganisms in the indoor air has remained outside the general topic of energy saving.

The papers included in this study collect a total of 8777 citations in WOS, of which 24.6% are concentrated in ten of them, as shown in Table 1. As can be seen, this generally involves research related to interior thermal comfort, the search of optimal thermal insulation and the use of renewable energy to improve the energy performance of houses.

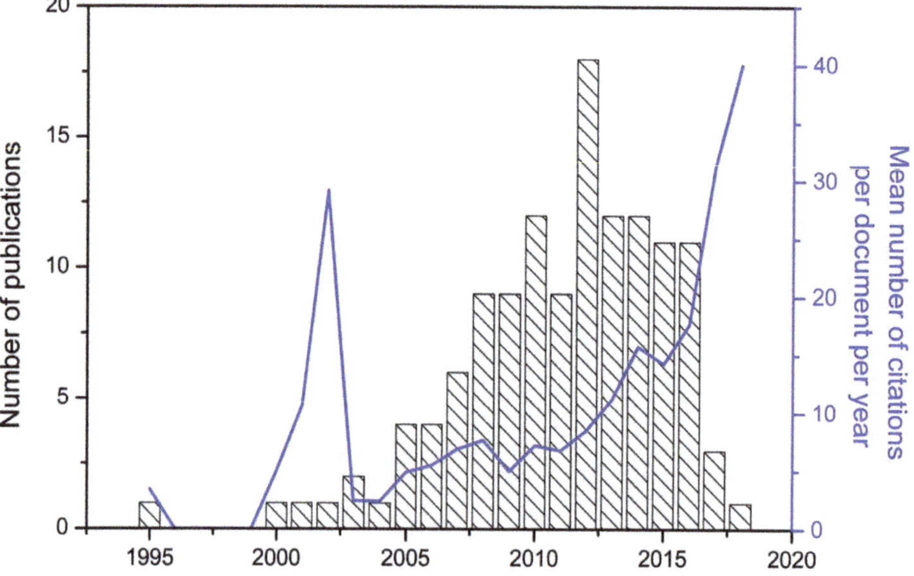

Figure 3. Temporal evolution of analysed publications.

The 128 publications analysed were produced by 359 authors. Table 2 presents the authors with the most significant impact, the total number of citations, the number of publications of each author used in this study and the main work of each author. Three-hundred-and-thirteen of the identified authors participated in only one publication, and the average citation value per author was 80. Most of the analysed studies are concentrated in China (24%) and the USA (20%), with the rest of the countries being in the minority (7% from Greece, 6% from Spain, 5% from Australia and the remaining 38% from 22 other countries); in fact, the five authors with the largest number of publications belong to Chinese institutions (Lam J.C, Yang L., Jing Y.Y, Li D.H.W. and Wang J.J.), which comprise 2847 of the 8777 citations.

Table 1. The most-cited bibliography used in the analysed researches.

Title of Publ.	Authors	Journal	Type	Year	Tot. Cit.	Tot. Cit./Year	Ref.
Thermal comfort in naturally ventilated buildings: revisions to ASHRAE Standard 55	De Dear R.J.; Brager G.S.	Energy and Buildings	Article proceedings paper	2002	496	29.2	[55]
Thermal comfort and building energy consumption implications—A review	Yang L.;Yan H.Y.; Lam J.C.	Applied Energy	Review	2014	308	61.6	[61]
Solar air conditioning in Europe—an overview	Balaras C.A.; Grossman G.; Henning H.M.; Ferreira C.A.I.; Podesser E.; Wang L.; Wiemken E.	Renewable & Sustainable Energy Reviews	Review	2007	241	20.1	[63]
Forty years of Fanger's model of thermal comfort: comfort for all?	Van Hoof J.	Indoor Air	Review	2008	220	20.0	[60]
The adaptive model of thermal comfort and energy conservation in the built environment	De Dear R.; Brager G.S.	International Journal of Biometeorology	Article	2001	192	10.7	[64]
Zero energy buildings and sustainable development implications—A review	Li D.H.W.; Yang L.; Lam J.C.	Energy	Review	2013	182	30.3	[65]
Determination of optimum insulation thickness for building walls with respect to various fuels and climate zones in Turkey	Bolatturk A.	Applied Thermal Engineering	Article	2006	150	11.6	[66]
Impact of climate change on energy use in the built environment in different climate zones—A review	Li D.H.W.; Yang L.; Lam J.C.	Energy	Review	2012	128	18.3	[67]
Energy performance of building envelopes in different climate zones in China	Yang L.; Lam J.C.; Tsang C.L.	Applied Energy	Article	2008	122	11.1	[68]
Extending air temperature setpoints: Simulated energy savings and design considerations for new and retrofit buildings	Hoyt T.; Arens E.; Zhang H.	Building and Environment	Article	2015	119	29.8	[69]

Table 2. Principal authors of the analysed researches.

Author	Last Affiliation (5 October 2019)	Num. Publ.	Sum. Cit.	Year of the First Publ.	Title of More Cited Publ.	Ref.
Lam J.C.	City University of Hong Kong, Kowloon, Hong Kong	8	983	2008	Thermal comfort and building energy consumption implications—A review	[61]
Yang L.	Xi'an University of Architecture and Technology, Xi'an, China	7	904	2008	Thermal comfort and building energy consumption implications—A review	[61]
Jing Y.Y.	North China Electric Power University, Baoding, Baoding, China	4	268	2010	Multi-criteria analysis of combined cooling, heating and power systems in different climate zones in China	[70]
Li D.H.W.	City University of Hong Kong, Kowloon, Hong Kong	4	424	2011	Zero energy buildings and sustainable development implications—A review	[65]
Wang J.J.	North China Electric Power University, Baoding, Baoding, China	4	268	2010	Multi-criteria analysis of combined cooling, heating and power systems in different climate zones in China	[70]
Balaras C.A.	National Observatory of Athens, Athens, Greece	3	379	2000	Solar air conditioning in Europe—an overview	[63]
De Dear R.	The University of Sydney, Sydney, Australia	3	326	2001	Thermal comfort in naturally ventilated buildings: revisions to ASHRAE standard 55	[55]
Hong T.Z.	Lawrence Berkeley National Laboratory, Berkeley, United States	3	158	2013	Commercial building energy saver: an energy retrofit analysis toolkit	[71]
Mago P.J.	Mississippi State University, Starkville, United States	3	137	2009	Analysis and optimisation of the use of CHP-ORC systems for small commercial buildings	[72]
Tsang C.L.	City University of Hong Kong, Kowloon, Hong Kong	3	251	2008	Energy performance of building envelopes in different climate zones in China	[68]
Ucar A.	Firat Üniversitesi, Elazig, Turkey	3	169	2009	Effect of fuel type on the optimum thickness of selected insulation materials for the four different climatic regions of Turkey	[73]
Wan K.K.W.	City University of Hong Kong, Kowloon, Hong Kong	3	195	2008	Building energy efficiency in different climates	[74]

3.2. Results of Cluster Analysis

Next, an analysis of the results of the Citation Network processing in Gephi, as well as the clustering, was carried out, identifying the main directions of scientific development in climate-oriented studies in building. As clustering was based on common references, Table 3 shows the analysis of the most commonly used references in the analysed studies. In total, 114 analysed studies cited 4172 publications. At the same time, work published in ASHRAE Transaction [75] is presented in the references of 12 analysed studies.

Table 3. Most used articles in the references of analysed studies.

Year	Journal/Book	Title of Publ.	Num. of Rep.	Ref.
1998	ASHRAE Transactions	Developing an adaptive model of thermal comfort and preference	12	[75]
2008	Applied Energy	Energy performance of building envelopes in different climate zones in China	9	[68]
1998	Energy and Buildings	Thermal adaptation in the built environment: a literature review	8	[76]
2011	Building and Environment	Future trends of building heating and cooling loads and energy consumption in different climates	8	[77]
2002	Energy and Buildings	Thermal comfort in naturally ventilated buildings: revisions to ASHRAE standard 55	7	[55]
1970	Book	Thermal comfort. Analysis and applications in environmental engineering.	7	[78]
2002	Energy and Buildings	Extension of the PMV model to non-air-conditioned buildings in warm climates	7	[79]
1998	ASHRAE Transactions	Understanding the adaptive approach to thermal comfort	7	[80]
2005	Renewable Energy	Energy policy and standard for built environment in China	7	[81]
2003	Applied Energy	Towards sustainable energy buildings	6	[82]
2009	Energy and Buildings	Analysis and optimisation of CCHP systems based on energy, economical and environmental considerations	6	[83]
2008	Energy and Buildings	Integration of distributed generation systems into generic types of commercial buildings in California	6	[84]
2012	Applied Energy	Impact of climate change on building energy use in different climate zones and mitigation and adaptation implications	6	[85]
2010	Applied Energy	Multi-criteria analysis of combined cooling, heating and power systems in different climate zones in China	6	[70]
2010	Energy	Environmental impact analysis of BCHP system in different climate zones in China	6	[86]
2011	Applied Energy	Influence analysis of building types and climate zones on energetic, economic and environmental performances of BCHP systems	6	[87]

The map of bibliographic coupling clusters obtained from Gephi analysis (Figure 4) made it possible to identify the following nine clusters, in addition to the general outline of the relationship between Citation Network works:

- Cluster 1—Mitigation of the effects of UHI and cooling of buildings (with the following principal topics: Building cooling, UHI mitigation techniques, Outdoor thermal comfort, General energy saving techniques and Indoor thermal comfort in UHI conditions).
- Cluster 2—Indoor air microorganisms.
- Cluster 3—Combined heating, cooling and power systems.
- Cluster 4—Economic and energy optimisation of the thermal insulation.
- Cluster 5—Indoor thermal comfort.
- Cluster 6—Energy optimisation of school buildings.
- Cluster 7—Infiltration and air leakage.
- Cluster 8—Windows and façade optimisation.
- Cluster 9—Energy simulation, conservation and meteorological data (with the following principal topics—Simulation of energy consumption, Multiparametric and multi-objective optimisations, Schedule occupancy optimisation, Heat and energy recovery ventilators, HVAC systems optimisation, Indoor thermal comfort models implementation for energy simulation, Climate change and energy consumption, Building climate zones and Meteorological data for simulation).

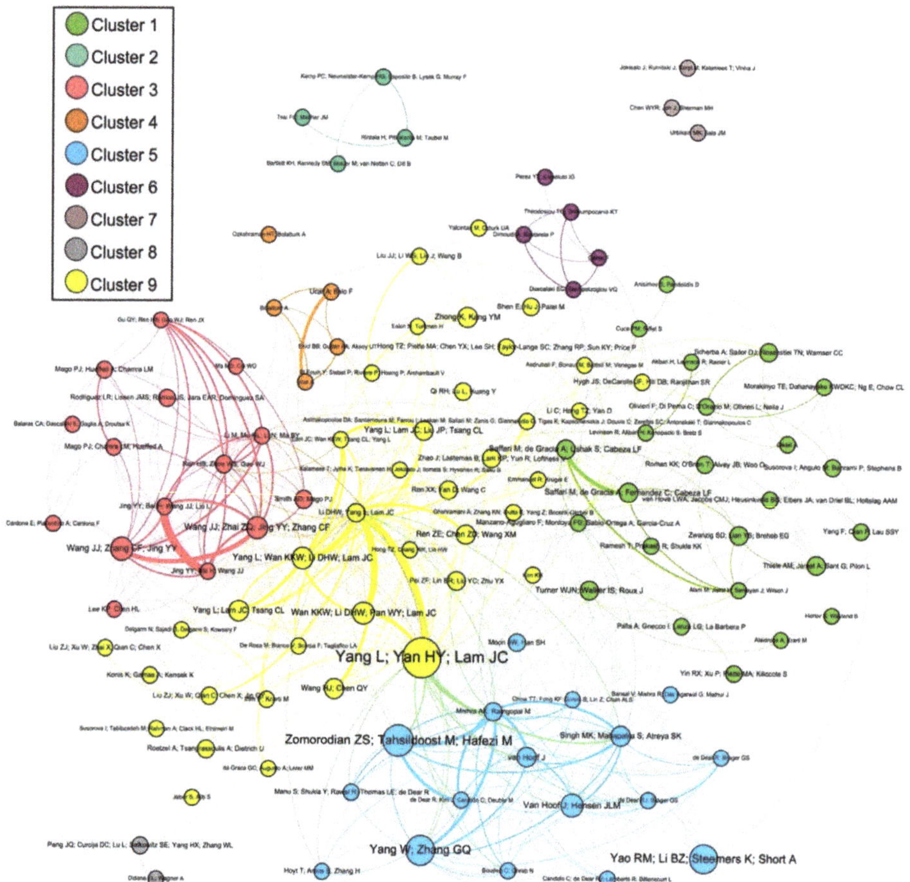

Figure 4. Map of bibliographic coupling clusters.

In this clustering, 114 publications have been used, with 517 interconnections (edges) of repeated references in each pair of papers, varying from one common reference to 41. Figure 4 shows that, of all the clusters identified, seven are connected to each other, and two are isolated; depending on the number of references in common, the thickness of the edge is greater if there are more references in common between two works. There is a greater number of common references in studies by the same authors, e.g., (Yang L; Yan HY; Lam JC) [61] and (Li DHW; Yang L; Lam JC) [67]. On the other hand, the size of the circle of each article depends on the degree of involvement of each work in the Citation Network of the study; thus, the work of (Yang L; Yan HY; Lam JC) [61] is grade 51, which means that 48 study network papers are used in his references, and three study network papers cite this work; instead, the research of (Ozkahraman HT; Bolatturk A) [88] is grade 1 and located on the periphery of the network, which means that only 1 study in the network uses this work in its references. Given that the present study used works with high citation numbers in the scientific environment, the main emphasis in the analysis of the results will focus on the clustering and the definition of the main themes of the studies, and not on how these works relate to each other in this research network.

In general, the seven clusters have a clear correspondence with one single scientific topic. However, Cluster 9 includes ten themes, due to its central position among other clusters; therefore, the themes of this cluster can overlap with the topics of other clusters. In addition, Cluster 1 includes five themes.

During the analysis of the publications, it was observed that most of the works are within an overall theme of energy conservation in the buildings. Outside the subject of energy conservation is the Indoor air microorganisms theme (Cluster 2). Moreover, the Indoor air microorganisms theme is isolated, just as most of the work of Indoor thermal comfort is outside the subject of energy conservation; this is because, although research has recently been carried out on the application of thermal comfort models in the process of energy simulation and for the evaluation of the energy efficiency of buildings, most of the studies in this cluster focused on the development of thermal comfort models. The topic of outdoor thermal comfort (Cluster 1) is also vaguely related to the topic of energy conservation. The economic and energy optimisation of the thermal insulation (Cluster 4) is difficult to relate completely to the topic of energy conservation, as the basis of the works on this theme is the methodology for finding an economic optimum between the insulation material and the economic costs of heating and cooling in buildings.

It can be said that the definition of clusters using Gephi and the bibliographic coupling method showed good results for identifying peripheral and isolated topics. However, in the case of centrally located topics and overlapping topics, additional manual analysis of the content of the clusters is required. Therefore, a detailed analysis of each cluster will be presented below.

3.2.1. Cluster 1—Mitigation of the Effects of UHI and Cooling of Buildings

The first cluster includes a total of 24 investigations (Table S1). These include general methodologies for energy conservation in buildings, techniques for mitigation the various effects of urban heat islands (UHI), the improvement of outdoor thermal comfort in UHI conditions and techniques for reducing energy consumption for cooling in buildings.

The first and oldest works of this cluster were focused on the idea of energy conservation. They analysed the application of cool-roof energy conservation techniques for non-residential and commercial buildings in different cities and in different climate zones, showing energy simulation results [58,59]. Subsequently, studies appeared that focus on the search for optimal bioclimatic architectural strategies [89] as well as the optimal design of residential building envelop systems [90], all of this with the general aim of reducing energy consumption in housing and improving indoor thermal comfort.

Mitigation techniques were developed, ranging from cool-roofs, green roofs and green façades, to the application of phase change materials (PCM) for building envelopes and the search of the optimal PCM for different cities and climate conditions. These generally show an evolution from field studies of the thermal effect and the radiation balance of various UHI mitigation techniques to complex computational simulations of the energy saving potential, considering indoor thermal comfort. Thus, in 2010, a paper was published on the search for optimal building shapes [91] with the main purpose to mitigate the effect of UHI using simulation with the ENVI-Met tool, which showed that the shape of buildings can improve ventilation in the streets, and the use of green-roofs, will help mitigate the effect of UHI. In this respect, research on techniques for mitigating the effect of UHI was subsequently carried out. Thus, in 2011, based on Energy Plus simulations in six ASHRAE climate zones in the USA, it was determined "that white and green-roofs are effective strategies for UHI mitigation" [92]. In Italy, based on field experiments and computer simulation, a numerical model for calculating the thermal resistance of green roofs was presented [93]; finally, numerical models were developed for calculating the thermal resistance of green façades [94]. Continuing with the application of UHI mitigation techniques in countries such as the USA and Australia [95], the use of PCMs in building envelopes, which can generate potential annual energy savings of 17–23%, is emerging [96]. In this regard, it has become clear that "PCM performance highly depends on the weather conditions, emphasising the necessity to choose different PCMs at different climate zones" [97]. More recent work focuses on finding optimal PCM melting temperatures for different cities and climate zones [98], as well as combining the use of PCM with indoor thermal comfort models in energy simulations [99].

This cluster also includes two publications that address the study of outdoor thermal comfort in UHI conditions, as well as the development of methods to mitigate the thermal and energy effects of UHI [100,101]. In both investigations, the authors point out the importance of the urban wind environment for outdoor thermal comfort in two cities located in different climate zones. By 2017, estimates were already being made of the effect of green roofs on outdoor/indoor temperature and cooling demand under four different climate conditions and urban densities with ENVI-met and Energy Plus tools [102]. In this sense, studies have evolved from focusing solely on the dwelling to an analysis of buildings with an aspect of interaction with the urban environment, e.g., on the scale of streets and neighbourhoods.

Additionally, because the effect of the increased energy consumption of buildings in UHI conditions is a problem generally in cities with hot climates and mostly related to cooling, this cluster presents research on energy reduction techniques for cooling. In 2010, two studies were published by different scientific groups in California on economic methods for energy conservation. In these cases, the critical-peak pricing effect [103] was studied in the case of cooling energy for residential housing. Furthermore, the optimisation of HVAC system operation settings (pre-cooling strategies) [104] for the minimisation of energy consumption in commercial buildings was analysed. In addition, the use of pre-cooling strategies was also used for residential buildings [105].

3.2.2. Cluster 2—Indoor Air Microorganisms

The second cluster, with four papers, deals with Indoor air microorganisms and includes studies on the presence of bacteria, fungi and microbes inside the office, school and residential buildings, as a consequence of the variability of these microorganisms, which is related to climatic zones and ventilation techniques, as well as solutions to reduce their presence.

Initially, airborne fungi species and culturable bacteria content analysis research focused on office buildings. In 2003, it was shown that an HVAC system helps to minimise the concentration of airborne fungi species in the indoor air of New York offices, whereas this is not the case in Perth [56]. In addition, the seasonal and climatic dependence on the content of culturable airborne bacteria in office indoor air of 100 office buildings in 10 US climate zones was revealed [57].

In another study, airborne bacterial concentration was analysed in 39 schools in Canada, showing a direct relationship between the concentration of airborne bacteria and the concentration of CO_2 in school classrooms, demonstrating the importance of implementing ventilation in school facilities in the climate zone studied [106]. Finally, in 2012, house dust microbial communities were investigated, and conclusions were drawn that residences located in a temperate climate zone showed higher dust microbial diversity than in tropical climate zone [107].

3.2.3. Cluster 3—Combined Heating, Cooling and Power Systems

The third cluster presents 15 investigations, listed in Table S2, dealing with climate-dependent optimisations and an analysis of energy saving potential of combined heating, cooling and power systems (CHCP).

Initially, the primary interest of the scientists was to investigate the energy effects of the use of CHCP systems in airport buildings due to a problem related to the high level of energy consumption in these types of buildings, while noting the good potential for energy savings from using these systems in more southerly climatic zones [108]. Thus, in the oldest study of the cluster, the problem of using systems with the right power in the aspect of energy conservation in airport buildings was raised. Systems optimisation measures at Greek airports could result in potential energy savings of 15–20% for the southern climates and 35–40% for the northern climates [109].

Subsequently, studies already focused on office buildings were addressed. Thus, in 2009, in the USA, a method was demonstrated to optimise a combined heating and power system (CHP) in this type of building, and in the context of three objectives: energy saving (ES), operation cost reduction (OCR) and environmental impact minimisation (EIM)—only for carbon emissions; it was shown that electricity and natural gas CHP system optimisation is dependent on climate zones and operational mode [110]. In the case of a building in Beijing, solar CHCP systems were considered, and the OCR optimisation objective was extended in the context of LCA methodology. Moreover, the EIM objective was extended to global warming, the chemical composition of precipitation and the respiratory effect of pollutants, and the results indicate that the energy saving and pollutant emissions reduction potentials of the FTL operation mode are better than that of the FEL mode [111]. The same authors of the previous study used a similar approach for multi-objective optimisation of the electricity and natural gas CHCP system in a commercial office building in Beijing. In this research, the optimal equipment size and operation mode was obtained. The authors found that "if energy benefits are paid more attention, following the thermal load (FTL) mode is the first choice, while if environment performance is more valuable, following the electricity load (FEL) mode is the good operation strategy" [112]. In the two papers presented above for Beijing, the authors note that their conclusions were obtained for the specific climatic conditions of this city.

In another investigation, the application of CHP systems with Organic Rankine Cycle (ORC) was analysed in small commercial office buildings in six climate zones in the USA. The authors show that the CHP–ORC system helps reduce the primary energy consumption cost and carbon dioxide emissions in comparison with the same building, operating solely with a CHP system in all climatic zones of study [72].

In the case of hotel buildings, optimisation methodologies aimed at three objectives (ES, OCR and EIM) were presented for CHCP systems. The authors concluded that CHCP systems are most relevant in the aspect of primary energy conservation and emission reduction in a cold climate with high heating energy needs [70]. On the other hand, in the USA, it proved effective in the hybrid method for operating a CHP system of a hotel building (which either follows the thermal or the electric demand), by comparing the energy simulation results of this method with FEL and FTL modes of operation strategies in 16 climate zones. It was shown that the choice of the CHP system operation method will depend on the climate zone for this type of building. Therefore, in all climatic zones, the efficiency of the system with the hybrid operational mode was greater than with FEL and FTL. However, the FEL operational mode is more suitable for warm climatic zones of ASHRAE (1A, 2A, 2B, 3A, and 3B) and FTL for cold climatic zones [113].

In the research of Wang et al., the authors developed and compared the use of optimal CHP systems for four types of buildings (hotel, school, office and hospital) in five cities in China, with different climates in the context of three objectives (ES, OCR and EIM). The authors concluded that energy demands depend on the type of building, which also affects the optimal design of the CHP. In general, throughout the year, the CHP system in office buildings consumes less energy, spends less and emits less CO_2 among the four categories of buildings. In terms of saving primary energy, CHP systems are optimal for severe cold climates; in terms of reducing carbon emissions, they are optimal for mild climates [87].

In 2012, the possibility of applying CHCP systems in a multi-storey residential building, under the climatic conditions of Shanghai City, was analysed. The authors summarised that, from an economic point of view, the introduction of these systems is not profitable for residential buildings and that a reduction in natural gas prices may make it possible to use these systems in such buildings [114]. On the other hand, in Spain, the design of the integrated hybrid solar thermal (PV) micro-CHP system was analysed and applied to an apartment building in five climate zones. In this paper, a discussion was presented on the problems of using micro-CHP systems in residential buildings, and it was concluded that CHP systems are optimal for cold climate zones [115].

The evolution of the application of CHCP systems for different types of buildings is remarkable. Initially, these systems were used for airport buildings and other types of public buildings. Recently, researchers have been analysing the possibility of applying these systems to residential buildings. In addition, it is significant that the timing and mode of operation of these systems depend on the type of building, climate zones and target (energy saving, economic or ecological) that is most desired with the use of CHCP systems.

3.2.4. Cluster 4—Economic and Energy Optimisation of the Thermal Insulation

In the fourth cluster, five studies are presented that develop a methodology for determining the optimum thickness of thermal insulation based on minimising the parameters that affect it.

Thus, the first paper presented a methodology for finding the optimal thickness of thermal insulation for exterior walls in a cold climate zone in Turkey. This methodology was based on an approach of minimising the economic costs of heating as well as the cost of thermal insulation [88]. Following the methodology of the previous study, subsequent work searched for the optimum thickness of the exterior wall insulation, considering the economic–energy balance using (i) heating degrees-days, (ii) the cost of the insulating material and (iii) the cost of fuel used to heat the house for a decade [66]. In subsequent studies, the methodology was extended by adding the search for the optimum fuel for different climate zones [73], as well as the search for the optimum thickness of the thermal insulation of the roof, evaluating also the effect of the thermal insulation on the environment, thus reducing CO_2 and SO_2 emissions [116] and, finally, extending the number of materials analysed for use as thermal insulation, determining, for each climate zone in Turkey, the optimum material and structure in the external walls [117].

3.2.5. Cluster 5—Indoor Thermal Comfort

The fifth cluster is composed of 17 papers on indoor thermal comfort, which generally deal with the concept of the adaptive thermal comfort model and its evolution over time, the conventional thermal comfort model and the potential for energy savings in the application of thermal comfort models in buildings. All studies of this cluster are presented in Table S3.

The adaptive thermal comfort (ATC) model for naturally ventilated buildings, developed by de Dear and Brager, as an alternative to the conventional model of thermal comfort [64], with all the limits of its application, the energy saving potential and the problems of its development [55], was adopted as an amendment to the ASHRAE 55 Standard and then went through different stages of evolution from its conceptual formation to its application in buildings to minimise energy consumption.

First, it was found that the ATC model has a clear dependence on local climatic conditions and the thermal preferences of the population in different regions of the world. Generally, the ATC model is more applicable to regions of the world with a warm climate [118–120]. These studies showed that the effect of indoor air speeds on thermal comfort in buildings with natural ventilation requires detailed research for different climate zones [119,120].

The application of the ATC model in colder climate conditions was also considered with estimates of potential energy savings [121]. It has been shown that the ATC model can only be used in the summer period to natural ventilated office space under Dutch climatic conditions, with a cooling energy saving potential of up to 10%. [122]. For conference and exhibition building in Beijing, a hybrid HVAC system operated with natural ventilation cooling and with the application of a high standard level of the ATC model can be used between 63% and 66% of the hours in July and August, and that can help to reduce energy consumption [121].

This cluster presents studies not only on the ATC model, but also on the conventional model of thermal comfort. Thus, in 2008, although the author accepts the best applicability of the ATC model for naturally ventilated buildings, van Hoof conducted a review on the predicted mean vote (PMV) model of thermal comfort created by Fangér, demonstrating its strengths and weaknesses. Possible methods of optimisation and improvement of the PMV model were also analysed [60]. Later, the authors of a study carried out in Hong Kong concluded on the importance of taking into account air velocity for thermal comfort in an air-conditioned office building and recommended the use of air-conditioning systems with individual control for this type of building [123].

The existence of two models of thermal comfort, of course, leads to the development of research in terms of comparing them with each other. For example, in India, for residential buildings, it was concluded that the ATC model is more suitable for reflecting the thermal preferences of the population than the PMV model [124]. A good comparative study of the two models was published in 2013 on field studies in thermal comfort research, presenting all research according to the four Köppen climate zones. The authors concluded that the conditioned spaces have narrower comfort zones compared to free-running buildings. In all climate zones, the most popular adaptation methods are related to the modification of air movement and clothing of the building occupants [125]. Additionally, there were still attempts to develop a single ATC model for air-conditioned and natural ventilated office buildings [126].

Most of the above studies focus on office and residential buildings. However, on the other hand, school buildings also receive significant attention from thermal comfort researchers. The interest in studying thermal comfort in schools is because children spend most of their time there, and their academic performance and emotional health depend on thermal comfort. In one of these works, the acceptable temperature range for Australian school children was analysed. This research was carried out in nine schools, in three climate zones, with different HVAC systems. It was identified that the comfortable thermal sensation of the children was between 1 and 2 °C lower than the thermal sensation of the adults. Within this research, a review of work on thermal comfort in schools and in different geographical areas is presented [127]. Furthermore, in different countries of the world, there are policies to optimise energy costs in educational institutions, which depend on indoor thermal comfort [128]. Generally, in previous investigations, it was concluded that thermal comfort depends on the type of building and the age and gender structure of the building users.

Following, after the stages of development and evolution of the models, other studies on the practical application of the thermal comfort models can be found, in which the adjustment of the thermostat configuration based on the thermal comfort models is discussed, to minimise the energy consumption of the buildings in different climatic zones [69,129]. Finally, in one study, the use of an earth–air–tunnel heat exchanger is analysed for the minimisation of the energy consumption maintaining thermal comfort, according to the ATC model of the ASHRAE standard [130]. It can be said that the choice of using an ATC model, a conventional model or a combination of both for the operation of an HVAC system to minimise energy consumption, depends on the climate zone in which the building is located.

3.2.6. Cluster 6—Energy Optimisation of School Buildings

The sixth cluster consists of five articles dealing with the energy optimisation of school buildings, mainly based on the objective of maintaining indoor comfort conditions and focusing on a holistic approach (energy efficiency, thermal comfort and indoor air quality).

This cluster addresses energy-efficient strategies in school buildings and different climate zones, based on aspects such as maximising energy conservation, improving indoor air quality and visual comfort [131]. The studies have shown that this requires optimising the design parameters of the building, its shape, orientation, thermal insulation and HVAC systems for different climate zones [132], light control, infiltration, glazing, night ventilation and size of windows [133], in addition to the use of renewable energy (usually solar), to minimise energy consumption in buildings [134,135].

3.2.7. Cluster 7—Infiltration and Air Leakage

The low number of works in this cluster, with only three publications, presents studies on the infiltration of residential buildings in general, as well as infiltration specifically of windows and exfiltration of dwellings. In this cluster, there are general conclusions on the need to develop studies on infiltration and exfiltration in different regions of the world for different types of housing and their effect on energy efficiency.

In Finland, the relationship between infiltration and the energy consumption of the heating and ventilation system of a single-family house was analysed. It was determined that infiltration causes 15–30% of the heating and ventilation energy use of the dwelling. In addition, the authors proposed a climate zoning methodology for infiltration [136]. In Spain, the effect of window infiltration on the process of energy certification of windows was analysed. Two energy balance equations through the window were presented for two climate zones in Spain, depending on window leakage at a pressure difference of 75 Pa and outdoor temperature difference of 20 °C. Based on the energy balance equation, it was possible to carry out an energy qualification of the windows [137]. The last work carried out in the USA analysed air leakage in 134,000 houses. It was established that the built year and climate zone are the two most influential parameters on air leakage. This is a good methodological work to understand the air leakage concept in relation to buildings [138].

3.2.8. Cluster 8—Windows and Façade Optimisation

The eighth cluster is the smallest of those analysed and is made up of two studies about the energy saving potential of semi-transparent photovoltaic panels for windows and façades. One of these presents a possible energy conservation evaluation methodology with semi-transparent PV windows in offices in two cities in Brazil and one in Germany [139]. The second one presents a methodology for finding the optimal design of a ventilated photovoltaic double-skin façade to minimise energy consumption in housing. [140]. In both cases, the energy saving efficiency of these systems was directly related to the climate zones.

3.2.9. Cluster 9—Energy Simulation, Conservation and Meteorological Data

This cluster, with 39 publications, is the largest, has a central location among the other clusters (Figure 4), and deals with energy conservation. All studies in this cluster are presented in Table S4 and include studies on the simplification of methodologies for estimating energy consumption in buildings [141,142]; the relationship between thermal comfort [61] and visual comfort [143] with energy conservation in buildings; and comparisons of the energy efficiency of simulated dwellings according to the use of national building standards, both in Italy and Spain, for different building climate zones, showing the most favourable energy regulations [144]. Due to its size, and to facilitate the analysis of the results, this cluster has been manually divided into five sub-themes, which will be analysed below: (i) Multiparametric and multi-objective optimisations; (ii) Heat and energy recovery ventilators; (iii) Schedule occupancy and occupant behaviour; (iv) Renewable energy systems; (v) Meteorological data and climate change.

Sub-Theme 1—Multiparametric and Multi-Objective Optimisations

This first sub-theme presents 13 studies on the optimisation of buildings for energy conservation through four different approaches in the use of energy-efficient measures (Table 4): (i) building envelopes, (ii) architectural parameters (building orientation and geometry), (iii) building service systems and equipment, and (iv) internal conditions (thermal/lighting comfort, building occupation, occupant behaviour and indoor environmental conditions).

Table 4. Principal studies of the multiparameter optimisations.

Parameters for Optimisation	Tools	Year	Ref.
Insulation and thermal mass; aspect ratio; the colour of ext. walls; glazing system; windows size; shading devices	Energy Plus and validation with measured data	2008	[145]
Shape coefficient; building envelops (wall, roof)	Num. model—Overall thermal transfer value (OTTV) method	2008	[68]
Passive solar design; internal loads (lightning and equipment); operations of fans and pumps; wall insulation	Simulation tool DOE-2.1 E	2008	[74]
Window design; 4 types of glazing	TRNSYS software Economic model life cycle cost (LCC) criterion	2011	[146]
Transparent composite façade system (TCFS) vs. glass curtain wall system (GCWS)	DOE-2 (eQuest) Economic model LCA	2011	[147]
Orientation; wall and roof ins.; win. size; WWR; glazing; lighting; infiltration; cooling Temp.; refrigerator energy effic. lev.; boiler type; cooling sist. type	DOE-2 LCC	2012	[148]
Win. orientation; WWR; width to depth ratio (W/D)	Energy Plus	2013	[149]
Building occupation; ATC model; CO_2 emission	Energy Plus	2014	[150]
6 principal parameters: climate; envelope; equipment; operation and maintenance; occupation behaviour; indoor environmental conditions	Energy use audit	2014	[151]
Energy conservation tool for optimising existing buildings and to design new buildings. 100 configurable energy conservation measures; IEQ	CBES toolkit Energy Plus	2015	[71]
Multicriteria optimisation; orientation; win. size; overhang specification	Multi-objective non-dominated sorting generic algorithm (NSGA-II) and Energy Plus	2016	[152]
Tool for multicriteria optimisation; passive environmental design strategies; building geometry; orientation; fenestration configuration y others.	Passive Performance Optimisation Framework (PPOF) Energy Plus	2016	[153]

It can be seen that the older publications considered a limited number of parameters for optimising buildings and only with an energy purpose. With the progress of research, the number of parameters and measures analysed has increased significantly, and secondary objectives of optimisation have also been developed, be they economic, ecological or health-oriented. These objectives were later integrated with the energy aspect of the optimisation. Recently, there has been a trend towards a decrease in the number of parameters and optimisation techniques, through the search for more efficient methods and parameters for different types of buildings, in different climate zones [65].

Sub-Theme 2—Heat and Energy Recovery Ventilators

This sub-theme presents three papers related to heat and energy recovery ventilators. In the case of heat recovery ventilators, in countries such as France and China, it has been shown that the energy efficiency of an optimal ventilator depends on the climate zone and the type of building [154,155]. Besides that, for an energy recovery ventilator, the seasonal dependence of weighted coefficients (latent and sensible heat efficient) of enthalpy efficiency in different cities was demonstrated [156]. These conclusions are derived from the direct dependence of the energy efficiency of these ventilators on the outside temperature, the moisture content of the air, and the latent and sensible heat content.

Sub-Theme 3—Schedule Occupancy and Occupant Behaviour

The third sub-topic includes three papers and focuses on assessing the impact of building occupancy on energy consumption and finding models of schedule occupancy closer to the real occupancy in residential [157] and office buildings [158]. In the case of offices, it was shown that the improvement of the schedule occupancy model led to a decrease in the energy consumption of the HVAC system during simulations in hot climate types and an increase in cold climate types. It also highlights the importance of selecting daily optimal setpoint temperatures for HVAC and demonstrates the potential for energy saving in different climates. For small, medium and large office buildings, setpoint temperatures for an HVAC system in the range of 22.5 ± 3 °C would lead to 10.09–37.03%, 11.43–21.01% and 6.78–11.34% energy saving, respectively, depending on the climate zones [159].

Sub-Theme 4—Renewable Energy Systems

The fourth sub-theme, with four papers, deals with energy saving techniques in buildings from the use of renewable energy-based systems such as solar thermal systems, photovoltaic systems [160], solar-assisted liquid desiccant air-conditioning systems [161], ground-source heat pumps [162] and hybrid ground-source heat pumps with an auxiliary heat source [163]. In this work, a clear dependence was shown on the possibilities and limits of the application of the above systems in the different climate zones.

Sub-Theme 5—Meteorological Data and Climate Change

This fifth sub-theme shows nine papers linking energy conservation and meteorological data. To simulate energy consumption, meteorological data is needed, so the quality and selection of the data are important.

Results of the use of typical meteorological year (TMY) data for energy simulation were compared with the use of real measurement data over a period of 30 years [164], with the use of typical principal component year (TPCY) data [165], and with the use of actual meteorological year (AMY) data [166]. It was concluded that, in the era of availability of meteorological data and the increasing power of computer technology, it is necessary to use real meteorological data from different years to simulate the energy consumption of households in order to obtain more realistic results or, as in Finland, to develop individual methodologies to create energy reference year data for energy simulations in different climate zones [167].

The simulation of energy consumption for the present, meanwhile, has generated scientific interest in the study of future energy consumption. This interest has a basis in the development of climate models and forecasts, as well as the unification of climate change projections. Thus, in 2011, researchers presented construction methods for adapting designed residential buildings to climate change [168]. In 2012, other authors pointed out the importance of the ability to analyse how buildings will respond to future climate changes and assessed possible quantitative changes in energy use and carbon footprints. In addition, they discussed possible climate change mitigation and adaptation strategies [85].

In the USA, changes in energy consumption in different types of buildings and climate zones were assessed according to IPCC climate change projections. It was found, by the 2080s, residential energy consumption will increase in climate zones 1–4 of ASHRAE and decrease in climate zones 6–7 of ASHRAE [25]. Similarly, in Greece, the effect of climate change on the energy consumption of buildings will be more noticeable in warmer climate zones [169].

3.3. Studies without Clusters

As noted above (Figure 2), 14 papers were left out of the clustering, which are presented in Table S5. In general, these studies are related to the topics that have been identified in the previous analysis of the clusters, such as thermal comfort and energy saving, CHCP and renewable energy systems, multicriterial optimisation for energy saving, climate change and change of energy consumption, etc.

Nevertheless, there are interesting researches, for example, on energy efficiency, where, with a group of analytical hierarchical processes, it was possible to identify 17 basic parameters out of 83 parameters that should be used to assess the energy efficiency of residential buildings in the hot summer and cold winter zones in China [170]. Likewise, studies are also presented on climatic building zones in Spain [171] and on the development of bioclimatic zones in India, with potential for the development of passive solar design strategies for residential buildings [172].

3.4. General Analysis of the Typology of Climate-Oriented Research

After a detailed description of the content of each cluster and a complete in-depth analysis of the papers in these clusters, it was concluded that there are seven main types of climate-oriented studies in building (I–VII). One-hundred-and-twenty of the 128 studies analysed could be classified into a particular type (Tables S1–S5). To facilitate an understanding of these seven types of studies, a schematic diagram is shown in Figure 5.

In the first macro-group, comparative works were classified. These were carried out to demonstrate differences between various climatic zones or geographical locations, as well as to develop general conclusions and patterns:

I. In 41.6% of the analysed studies, the research involves studies in a single geographical location for each climate zone studied. The results and conclusions obtained for each geographical point are extrapolated to the whole climate zone, and this logic applies to all climate zones of the study. The results for the climate zones are then compared with each other or are used for developing general conclusions.

II. In 26.7% of the studies analysed, the research involves studies in more than one geographic location for each climate zone studied. The results and conclusions obtained for different geographical locations are extrapolated to the whole climate zone, and this logic applies to all climate zones of the study. The results for climate zones are then compared with each other or are used to develop general conclusions.

III. A small proportion (6.7%) of the studies analysed compares results in different geographical locations to obtain general conclusions, without focusing on climatic zones. In the second macro-group are investigations on the development of recommendations, standards, and conclusions for a specific area or geographical location:

IV. In 10% of the studies analysed, conclusions were drawn for a climatic zone, extrapolating results and conclusions from a representative geographical location.

V. In 5.8% of the analysed studies, conclusions were drawn for one climate zone, extrapolating results and conclusions from several representative geographical locations.

VI. A small proportion (4.2%) of the studies analysed were designed to obtain conclusions and results for a specific geographical location, without focusing on climatic zones. Finally, the last group of studies:

VII. A small proportion (5%) of the studies analysed focused on the development of climate zones, including the analysis and identification of urban climate zones.

Figure 6a shows the time distribution graph of the main types of climate-oriented research. It can be observed that, in general, up to 2006, comparative works of Types I, II and III prevailed. After 2006, studies of Types IV, V, VI and VII started to appear. The maximum variability of climate-oriented research type in building was in 2010. This may be connected to the logical feasibility of doing comparative work first, looking for common patterns and differences in results in order to develop conclusions. This type of work leads to the need to generate studies to develop and establish standards and techniques, to find optimal systems and their correct functioning for certain geographical points and/or climate zones. For this reason, the studies corresponding to Types IV, V and VI appeared later. Similarly, there is a need for works on the development of climate zones; as these are more specific investigations, they require more involvement from specialists in different areas of science and the use of multiple climatological and meteorological data. Finally, these studies require more time and computing resources. Actual examples of Type VII are studies on the definition of new climatic zones for building construction [173,174] and urban local climate zones [175].

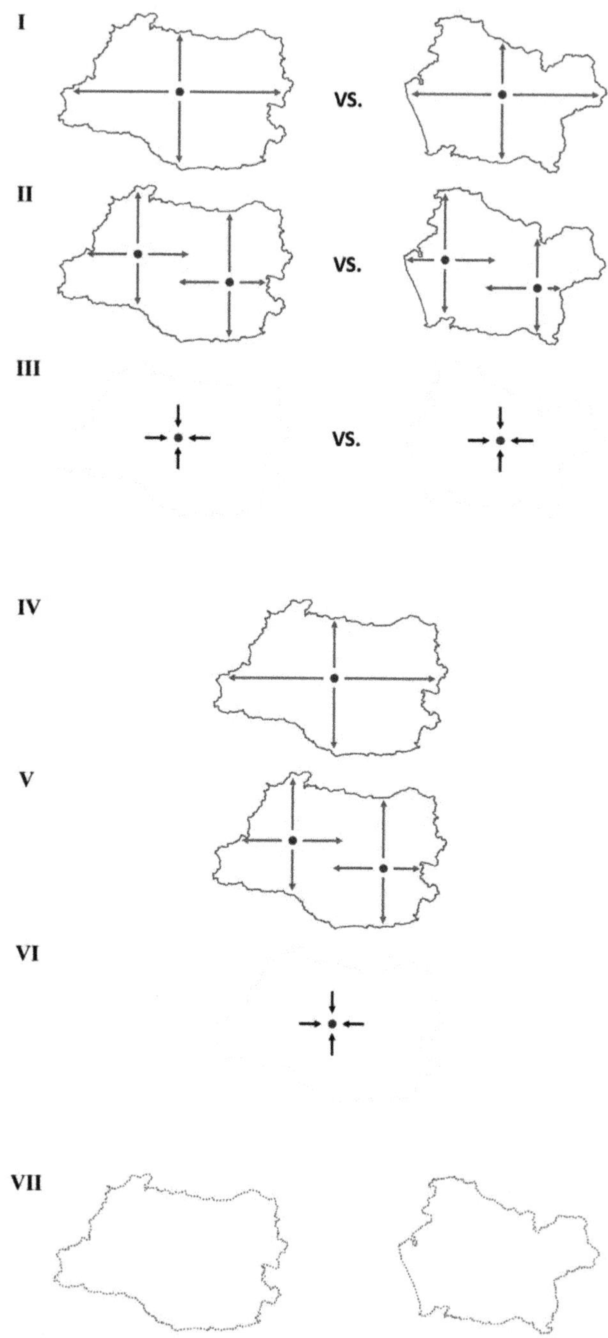

Figure 5. Seven main types of climate-oriented researches in building.

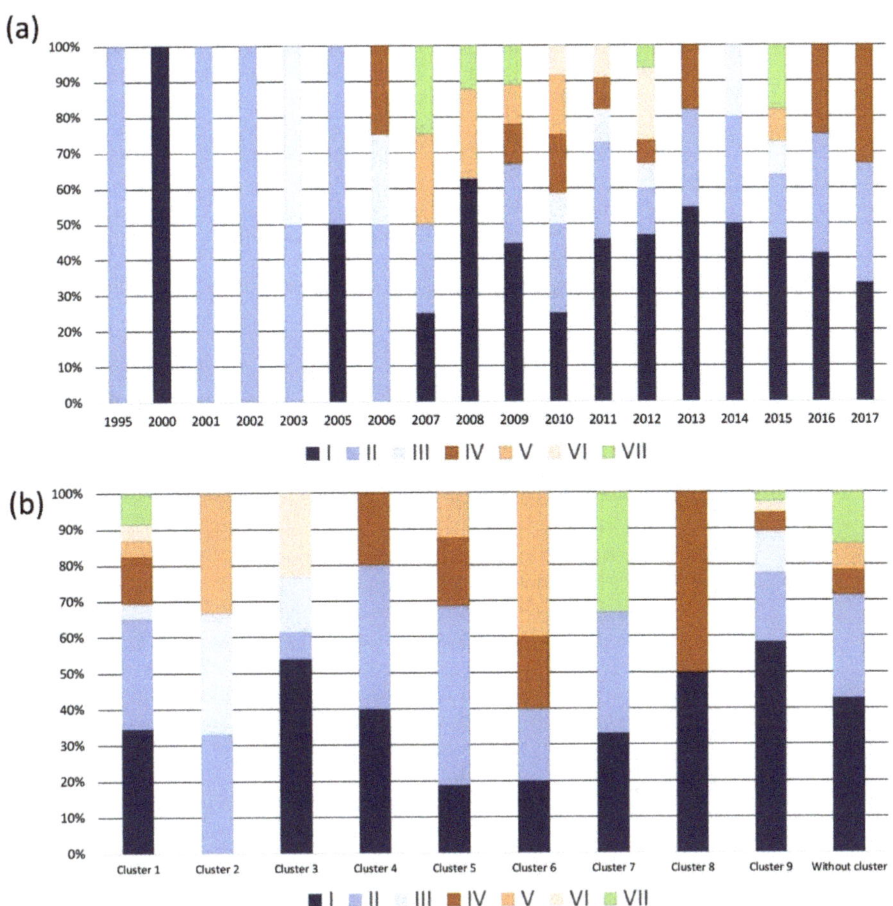

Figure 6. Temporal distribution of the seven main types (**a**) and content of each of them in the nine clusters analysed (**b**).

Figure 6b shows the remarkable difference in variability of the types of climate-oriented research in the clusters. Additionally, the types of analysed studies are presented in Tables S1–S5. Cluster 1 presents the largest variability in the types of climate-oriented research, from research on the establishment of urban climate zones, to the qualification of the application of UHI mitigation techniques for certain specific geographical locations. Among the other clusters with a large number of studies, Clusters 3 and 9 are mostly characterised as comparative studies (>75%) and Type I (>50%). This is due to the particularities of this works, which are based on the use of meteorological data for a single reference point, extrapolation of the results to a climatic zone and comparison with other zones. In addition, Cluster 3 has a higher number of Type VI studies, which corresponds to research carried out in a single geographical location without focusing on the climatic zone. In comparison, Cluster 5 has 32% of the Type IV and V studies, generally associated with the development of ATC models for different climate zones and specific regions based on a single or several geographical locations.

In general, it can be concluded that among all the investigations analysed, those corresponding to Type I are more representative, so concerns arise about the adequacy of the extrapolation of the results and the local conclusions for the whole climate zone.

Certainly, the practice of using one geographical point to cover a specific climate zone should no longer be used for climate-oriented research in building. Type II and V studies must be carried out in a more frequent manner. For example, recently, Type V research related to the assessment of the energy demand of buildings in various countries was published [176,177]. Additionally, recent work corresponding to Type II about the procedure for selecting glazing at different points of the same bioclimatic zone has been published [178].

Furthermore, the typology presented for climate-oriented researches in building can be applied and used in future research, not only in the scientific area of buildings.

3.5. General Analysis of Types of Climate Zoning and Climate Zones Used in Climate-Oriented Research

As climate-oriented research was analysed, the main types of climate zoning used in the reviewed studies are presented. In addition, the climate zones most commonly engaged in the research were analysed. Figure 7 shows an overview of the different types of climate zonings analysed in this study. It can be seen that most of the works included climate zonings from the National Building Codes (BCs). In 120 studies, where it was possible to identify the type of zoning, the National BC climate zonings were used 74 times. The total number of times zonings were used exceeds the number of studies, because some research has included more than one type of climate zoning. Additionally, the types of zoning and principal zones of the studied research are presented in Tables S1–S5.

Among the most popular National BC climate zonings were China (used 17 times in the studies reviewed); the USA, specifically ASHRAE zones (15); and Turkey (8). It can be noted that in the USA, some research has included the use of climate zones of the International Energy Conservation Code (IECC), the California Energy Commission (EC), the U.S. Energy Information Administration (EIA) and the Statewide Pricing Pilot (SPP) and climate zones of the California Public Utilities Commission. The IECC zones were generally used for studies related to energy analysis. The California EC zones were applied for state-oriented work in California.

Within the climate zones of National BC, China's "hot summer and cold winter" climate zone was the most popular in the works analysed. This zone includes the cities of Shanghai, Nanjing, Nanchang, Wuhan and Hefei, among others. Beijing, in contrast, is in the "cold" zone. Note that Chinese studies show a clear tendency to use representative geographical locations for each climate zone. China's research usually attempts to cover the whole country, using at least one representative location for each of its five climatic building zones.

In the case of the USA and the ASHRAE zones, the climate zones most commonly used in the studies are 3B and 4A. Zone 3B represents the cities of Los Angeles, San Diego and Sacramento, among others. On the other hand, zone 4A represents the cities of New York, Washington and Philadelphia, among others. Zones 1A, 1B and 5C are less represented. Moreover, there are no works with climate zone 0. This is related to the climatic reality of the USA, as there are no such zones in the territory of this country. For this reason, for some studies, geographical points from other countries of the world are used to cover the areas that are missing. Additionally, note that due to the clear ASHRAE zoning methodology, it is possible to identify ASHRAE zones in other geographical locations around the world. Among all the studies analysed in this research with National BC zones, only ASHRAE zones have this characteristic. For example, recent work on the analysis of the applicability of ASHRAE climate zones in the territory of China [179].

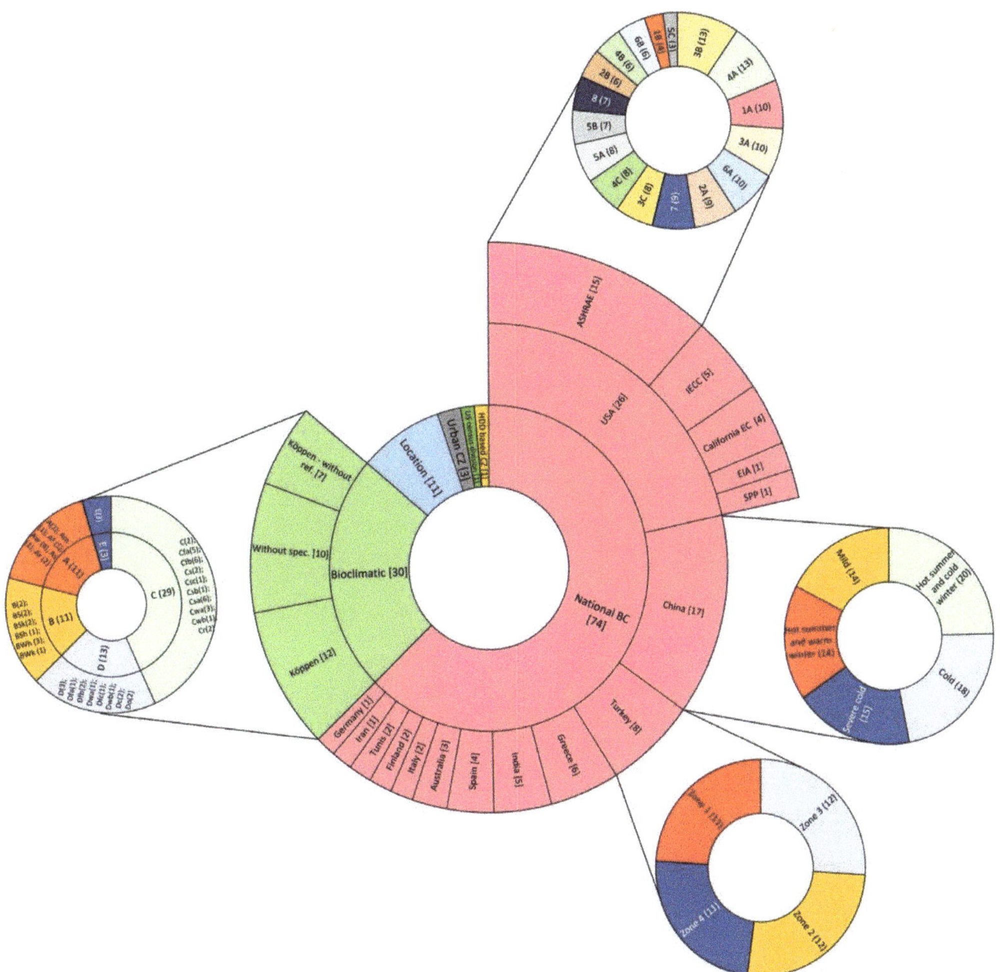

Figure 7. Main climate zonings [number of times of use] in analysed works and the most used climate zones (the number of geographical points corresponding to each zone).

In the case of Turkey, almost all the works generally involve the four climate zones of Turkish Thermal Insulation Standard TS 825. In this standard, zone 1 represents cities with warmer thermal conditions—Antalya and Izmir, among others. Zone 4 represents cities with colder thermal conditions—Erzurum and Kars, among others. Istanbul and Ankara are categorised in zones 2 and 3, respectively. The papers corresponding to Cluster 4 were totally realised under Turkish climate zones. At the same time, the methodology of this cluster has firmly entered the scientific field of building construction, which is reflected in its use for actual research [180,181].

Other National BC climate zonings exist: for example, in studies in Greece, climate zones from buildings' Thermal Insulation Regulation (TIR) and the Hellenic Regulation on the Energy Performance of Buildings (REPB) were used. Mostly these zones have been used in the research grouped in Cluster 6. The climate zones of Energy Conservation BC of India were used in three investigations belonging to cluster 5, "Indoor thermal comfort".

Moreover, in the climate-oriented studies, bioclimatic zones were used 30 times, generally, from the Köppen or Köppen–Geiger classification. The papers reviewed did not necessarily include bibliographical references that would allow the type of classification to be identified. In certain cases, it was difficult to establish the type of classification, because a quantitative description of the climate was used, for example, "warm climate" or "temperate climate". However, among the works where the zones could be identified, the most used Köppen zones in the climate-oriented works analysed were identified (Figure 7). It can be seen that climate type C (Temperate climates) is the most popular, analysed in 29 geographical locations, where Cfb (Oceanic climate or Marine west coast climate) and Csa (Mediterranean hot summer climates) are the Köppen subzones most commonly used. Zone Cfb represents cities located on the Atlantic coast of northwest Europe, southwest Australia and New Zealand. Zone Csa represents cities located in the Mediterranean Sea region, part of the Pacific coast of the USA and part of Australia. Additionally, it can be said that Köppen's climate zone E (Polar and alpine climates) is less represented in the climate-oriented building studies analysed. On the other hand, 11 studies involved specific geographical locations, three of which dealt with urban climate zones (Figure 7).

The distribution of climate zoning types used in the different clusters is presented in Figure 8. It can be seen that among the clusters with a large number of investigations, i.e., Clusters 5 and 1, there is a higher percentage of work based on bioclimatic zones, compared to Clusters 3 and 9, plus the "without clusters" investigations, which are related to energy simulations based on National BC climate zones. This is because, in the case of simulation-related research, building regulations strictly related to climatic building zones must be applied, as, for example, in current research of the assessment of the profile of the use of the HVAC system based on the Spanish national BC in the aspect of energy efficiency of residential buildings [182]. In the case of thermal comfort studies, and urban areas studies, outdoor comfort works can be carried out using bioclimatic zones, such as, for example, in actual studies about adaptive comfort in temperate and tropical climates [183,184].

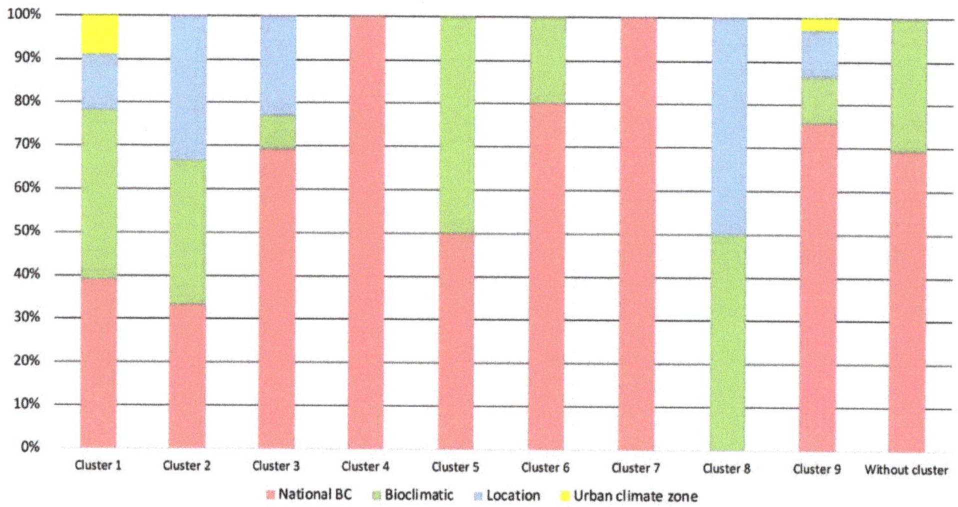

Figure 8. Main climate zonings for the clusters.

Note that most of the studies have a well-determined local view, these were carried out in climate zones established under the National BCs of different countries. In general, it can be said that differences were revealed both in the typologies of studies and in the types of climate zoning used in the works of various clusters and topics.

3.6. Future Lines of Research and Recommendations

The following are the main recommendations for future climate-oriented studies, both in the scientific field and in decision-making in public administration in building, considering the trends and knowledge gaps observed:

- In issues connected with UHI, it was found that there is a need to develop research and analyse the effect of UHI on energy consumption for the heating and cooling of dwellings in regions with cold climatic conditions and on climatic zoning for building.
- The topic of Indoor air microorganisms should be integrated with that of thermal comfort, and IEQ control should be applied to different types of buildings.
- In the area of thermal comfort, it is important to carry out research on the development of adaptive thermal comfort models for different age groups, for different types of buildings and in different geographical areas, and to a greater extent applicate in the practice of the thermal comfort models for functioning HVAC systems.
- As far as visual comfort is concerned, it is important to carry out studies on different types of buildings and climate zones. The use of visual comfort is also an objective in the multi-objective energy optimisation of school buildings.
- In terms of different building systems, it is important to develop comparative work in search of optimal parameters and of operating modes for CHCP systems for different types of buildings in various climate zones, to analyse hybrid building HVAC systems and energy saving capabilities in different climate zones; study possibilities for the integration of renewable energy systems; and analyse and map renewable energy capacities, which can be used for air conditioning, domestic hot water, and electrical energy in buildings in different regions of the world.
- In climate zoning, the need for research on dynamic building standards, that consider the effects of future climate change is notable. On the other hand, it is important to carry out studies on the climatic zoning of infiltration and to analyse further the energy effects of infiltration in different regions of the world.

It is possible to implement all the recommendations presented above under actual climatic conditions and taking into account the effect of climate change in future periods. In general, a significant amount of research related to climate change, the dynamics of climatic zones and modifications in meteorological parameters can be implemented. At the same time, note that the Sixth Assessment Report (AR6) of the IPCC is expected to be published in the coming years. This document will present new climate change assumptions, which will be based on four geophysical RCP scenarios of the AR5 and five new Shared Socio-Pathways (SSP) scenarios [185]. The development of Phase 6 of the Coupled Model Intercomparison Project (CMIP6) is also expected [186]. Approximately 70 climate modelling centres are involved in this project. The quality of climate modelling is expected to improve significantly. So far, a small number of climate modelling centres have completed the calculations. The first studies in the field of energy using new climate change scenarios have already been published [187,188]. Therefore, it is recommended that the possibility of using new SSP scenarios and climate modelling data from CMIP6 in future research in building be noted, as most studies are now carried out considering the effect of climate change.

4. Conclusions

The main conclusions of a bibliographic and manual analysis carried out on the climate-oriented publications with the highest impact in building are presented below.

The bibliographic coupling clustering method demonstrated reliable results in the scientific map definition, which was validated with manual analysis. According to that, a broad spectrum of issues directly and indirectly related to climate and climatic zones in building was defined. The climatic factor was identified in many aspects of the building construction—not only in studies on building climate zones and on the analysis of the effects of climate change, which represent only 3.1% and 4.7% of all studies, respectively, but also in studies on the optimisation of various building systems in different climate zones.

Eighty-eight percent of all climate-oriented studies, to one degree or another, are within the scope of the overall topic of energy conservation. Understanding the fact that energy conservation is directly or indirectly related to climate and climatic zones in buildings will help specialists in architecture and civil engineering design climate-appropriate housing in various aspects, for different regions of the world. In other words, it will be helpful to consider, at the design stage, many parameters, techniques and methodologies of buildings influenced by the climate factor. Additionally, the possibilities for the adaptation and mitigation of buildings in the context of climate change can be maximised.

Furthermore, this work is a significant contribution, as it presents a methodology that facilitates the establishment of a typology of climate-oriented studies that can be used in future research and is not only related to buildings. Variations in the types of research were observed in different clusters and topics. At the same time, the predominant type of studies analysed (41.6%) was comparative studies, carried out for different climate zones, with one representative geographical location in each zone.

From the point of view of the specific climate zones in which the analysed investigations were carried out, the prevalence of National BC climate zones and bioclimatic zones was revealed. It was found that, in the studies directly related to the simulation of the energy consumption of dwellings, research on the National BC climate zones is prevalent. In the case of studies focusing on indoor and outdoor thermal comfort, a significant proportion of them were carried out in bioclimatic zones. One-third of the research analysed was carried out in the climatic building zones of three countries—China, the USA and Turkey. The prevailing types of climatic zones correspond to regions with a large population. In addition, it was found that the only National BC climate zoning that was used in other countries outside the USA was ASHRAE.

Finally, note that the results obtained from the present study will help researchers to find knowledge gaps and identify new opportunities for studies connected to climate and climatic zones in building. In addition, this work can be used in the public administration field for the development and updating of building codes in different countries.

Supplementary Materials: The following are available online at https://www.mdpi.com/article/10.3390/app11073251/s1, Table S1: Studies of Cluster 1; Table S2: Studies of Cluster 3; Table S3: Studies of Cluster 4; Table S4: Studies of Cluster 9; Table S5: Studies without cluster.

Author Contributions: Conceptualisation, K.V., M.Z., C.S.-C. and M.C.; methodology, K.V. and C.S.-C.; funding acquisition, M.C.; software, C.S.-C.; validation, K.V. and C.S.-C.; formal analysis, K.V. and M.Z.; investigation, K.V., M.Z., C.S.-C. and M.C.; resources, C.S.-C. and M.C.; writing—original draft preparation, K.V. and M.C.; writing—review and editing, K.V., M.Z., C.S.-C. and M.C.; supervision, M.Z., C.S.-C. and M.C. All authors have read and agreed to the published version of the manuscript.

Funding: This research was supported by the Agencia Nacional de Investigación y Desarrollo (ANID) of Chile, through the projects: ANID FONDECYT 1201052; ANID PFCHA/DOCTORADO BECAS CHILE/2019—21191227; and research group TEP-968 Tecnologías para la Economía Circular of the University of Granada, Spain.

Institutional Review Board Statement: Not applicable.

Informed Consent Statement: Not applicable.

Data Availability Statement: Data are provided upon request to the corresponding author.

Conflicts of Interest: The authors declare no conflict of interest.

Nomenclature

AMY	Actual meteorological year
ASHRAE	The American Society of Heating, Refrigerating and Air-Conditioning Engineers
ATC	Adaptive thermal comfort
BC	Building code
CDD	Cooling degree-days
CHCP	Combined heating, cooling and power system
CHP	Combined heating and power system
CMIP6	Coupled Model Intercomparison Project (v 6)
CZ	Climate zone
EIM	Environmental impact minimisation
ES	Energy saving
FEL	Following the electricity load operation strategy
FTL	Following the thermal load operation strategy
GHG	Greenhouse gases
HDD	Heating degree-days
HRV	Heat recovery ventilator
HVAC	Heating, ventilation and air conditioning systems
IEQ	Indoor environmental quality
IPCC	Intergovernmental Panel on Climate Change
LCA	Life cycle assessment
LCC	Life cycle cost
OCR	Operation cost reduction
PCM	Phase change materials
PMV	Predicted mean vote thermal comfort model
PV	Photovoltaics systems
RCP	Representative Concentration Pathway
SSP	Shared Socio-Economic Pathway
TMY	Typical meteorological year
TPCY	Typical principal component year
UHI	Urban heat island
WOS	Web of Science
WWR	Windows/wall ratio

References

1. Hossain, M.U.; Ng, S.T. Strategies for enhancing the accuracy of evaluation and sustainability performance of building. *J. Environ. Manag.* **2020**, *261*, 110230. [CrossRef] [PubMed]
2. de Klijn-Chevalerias, M.; Javed, S. The Dutch approach for assessing and reducing environmental impacts of building materials. *Build. Environ.* **2017**, *111*, 147–159. [CrossRef]
3. Marique, A.-F.; Rossi, B. Cradle-to-grave life-cycle assessment within the built environment: Comparison between the refurbishment and the complete reconstruction of an office building in Belgium. *J. Environ. Manag.* **2018**, *224*, 396–405. [CrossRef]
4. Pal, S.K.; Takano, A.; Alanne, K.; Siren, K. A life cycle approach to optimizing carbon footprint and costs of a residential building. *Build. Environ.* **2017**, *123*, 146–162. [CrossRef]
5. UNEP's (SBCI). *Buildings and Climate Change: Summary for Decision-Makers*; UNEP: Nairobi, Kenya, 2009; ISBN 978-9280730647.
6. Peng, C. Calculation of a building's life cycle carbon emissions based on Ecotect and building information modeling. *J. Clean. Prod.* **2016**, *112*, 453–465. [CrossRef]
7. Liu, G.; Chen, R.; Xu, P.; Fu, Y.; Mao, C.; Hong, J. Real-time carbon emission monitoring in prefabricated construction. *Autom. Constr.* **2020**, *110*, 102945. [CrossRef]
8. Pérez-Lombard, L.; Ortiz, J.; Pout, C. A review on buildings energy consumption information. *Energy Build.* **2008**, *40*, 394–398. [CrossRef]
9. Pomponi, F.; Moncaster, A. Embodied carbon mitigation and reduction in the built environment—What does the evidence say? *J. Environ. Manag.* **2016**, *181*, 687–700. [CrossRef]
10. Eguía Oller, P.; Alonso Rodríguez, J.M.; Saavedra González, Á.; Arce Fariña, E.; Granada Álvarez, E. Improving the calibration of building simulation with interpolated weather datasets. *Renew. Energy* **2018**, *122*, 608–618. [CrossRef]
11. Silvero, F.; Lops, C.; Montelpare, S.; Rodrigues, F. Generation and assessment of local climatic data from numerical meteorological codes for calibration of building energy models. *Energy Build.* **2019**, *188–189*, 25–45. [CrossRef]

12. Bai, L.; Wang, S. Definition of new thermal climate zones for building energy efficiency response to the climate change during the past decades in China. *Energy* **2019**, *170*, 709–719. [CrossRef]

13. Verichev, K.; Carpio, M. Climatic zoning for building construction in a temperate climate of Chile. *Sustain. Cities Soc.* **2018**, *40*, 352–364. [CrossRef]

14. Xiong, J.; Yao, R.; Grimmond, S.; Zhang, Q.; Li, B. A hierarchical climatic zoning method for energy efficient building design applied in the region with diverse climate characteristics. *Energy Build.* **2019**, *186*, 355–367. [CrossRef]

15. Walsh, A.; Cóstola, D.; Labaki, L.C. Performance-based validation of climatic zoning for building energy efficiency applications. *Appl. Energy* **2018**, *212*, 416–427. [CrossRef]

16. Praene, J.P.; Malet-Damour, B.; Radanielina, M.H.; Fontaine, L.; Riviere, G. GIS-based approach to identify climatic zoning: A hierarchical clustering on principal component analysis. *Build. Environ.* **2019**, *164*. [CrossRef]

17. Yang, X.; Yao, L.; Peng, L.L.H.; Jiang, Z.; Jin, T.; Zhao, L. Evaluation of a diagnostic equation for the daily maximum urban heat island intensity and its application to building energy simulations. *Energy Build.* **2019**, *193*, 160–173. [CrossRef]

18. Toparlar, Y.; Blocken, B.; Maiheu, B.; van Heijst, G.J.F. Impact of urban microclimate on summertime building cooling demand: A parametric analysis for Antwerp, Belgium. *Appl. Energy* **2018**, *228*, 852–872. [CrossRef]

19. Bahramian, M.; Yetilmezsoy, K. Life cycle assessment of the building industry: An overview of two decades of research (1995–2018). *Energy Build.* **2020**, *219*, 109917. [CrossRef]

20. Aslam, M.S.; Huang, B.; Cui, L. Review of construction and demolition waste management in China and USA. *J. Environ. Manage.* **2020**, *264*, 110445. [CrossRef]

21. Gan, V.; Lo, I.; Ma, J.; Tse, K.; Cheng, J.; Chan, C. Simulation optimisation towards energy efficient green buildings: Current status and future trends. *J. Clean. Prod.* **2020**, *254*, 120012. [CrossRef]

22. Zhai, Z.J.; Helman, J.M. Implications of climate changes to building energy and design. *Sustain. Cities Soc.* **2019**, *44*, 511–519. [CrossRef]

23. Verichev, K.; Zamorano, M.; Carpio, M. Effects of climate change on variations in climatic zones and heating energy consumption of residential buildings in the southern Chile. *Energy Build.* **2020**, 109874. [CrossRef]

24. Wang, X.; Chen, D.; Ren, Z. Assessment of climate change impact on residential building heating and cooling energy requirement in Australia. *Build. Environ.* **2010**, *45*, 1663–1682. [CrossRef]

25. Wang, H.; Chen, Q. Impact of climate change heating and cooling energy use in buildings in the United States. *Energy Build.* **2014**, *82*, 428–436. [CrossRef]

26. Dolinar, M.; Vidrih, B.; Kajfež-Bogataj, L.; Medved, S. Predicted changes in energy demands for heating and cooling due to climate change. *Phys. Chem. Earth Parts A/B/C* **2010**, *35*, 100–106. [CrossRef]

27. Jylhä, K.; Jokisalo, J.; Ruosteenoja, K.; Pilli-Sihvola, K.; Kalamees, T.; Seitola, T.; Mäkelä, H.M.; Hyvönen, R.; Laapas, M.; Drebs, A. Energy demand for the heating and cooling of residential houses in Finland in a changing climate. *Energy Build.* **2015**, *99*, 104–116. [CrossRef]

28. Rey-Hernández, J.M.; Yousif, C.; Gatt, D.; Velasco-Gómez, E.; San José-Alonso, J.; Rey-Martínez, F.J. Modelling the long-term effect of climate change on a zero energy and carbon dioxide building through energy efficiency and renewables. *Energy Build.* **2018**, *174*, 85–96. [CrossRef]

29. Eshraghi, H.; Ansari, M.; Moshari, S.; Gholami, J. Climatic zoning and per capita demand forecast of Iran using degree-day method. *Adv. Build. Energy Res.* **2019**, 1–26. [CrossRef]

30. Ramon, D.; Allacker, K.; De Troyer, F.; Wouters, H.; van Lipzig, N.P.M. Future heating and cooling degree days for Belgium under a high-end climate change scenario. *Energy Build.* **2020**, *216*, 109935. [CrossRef]

31. Far, C.; Far, H. Improving energy efficiency of existing residential buildings using effective thermal retrofit of building envelope. *Indoor Built Environ.* **2018**, *28*, 744–760. [CrossRef]

32. Silvero, F.; Lops, C.; Montelpare, S.; Rodrigues, F. Impact assessment of climate change on buildings in Paraguay—Overheating risk under different future climate scenarios. *Build. Simul.* **2019**, *12*, 943–960. [CrossRef]

33. Bollo, C.S.; Cole, R.J. Decoupling climate-policy objectives and mechanisms to reduce fragmentation. *Build. Res. Inf.* **2019**, *47*, 219–233. [CrossRef]

34. Geng, A.; Chen, J.; Yang, H. Assessing the Greenhouse Gas Mitigation Potential of Harvested Wood Products Substitution in China. *Environ. Sci. Technol.* **2019**, *53*, 1732–1740. [CrossRef] [PubMed]

35. Deetjen, T.A.; Conger, J.P.; Leibowicz, B.D.; Webber, M.E. Review of climate action plans in 29 major U.S. cities: Comparing current policies to research recommendations. *Sustain. Cities Soc.* **2018**, *41*, 711–727. [CrossRef]

36. Graham, P.; Rawal, R. Achieving the 2 °C goal: The potential of India's building sector. *Build. Res. Inf.* **2019**, *47*, 108–122. [CrossRef]

37. Thonipara, A.; Runst, P.; Ochsner, C.; Bizer, K. Energy efficiency of residential buildings in the European Union—An exploratory analysis of cross-country consumption patterns. *Energy Policy* **2019**, *129*, 1156–1167. [CrossRef]

38. Kazemi, F.; Mohorko, R. Review on the roles and effects of growing media on plant performance in green roofs in world climates. *Urban For. Urban Green.* **2017**, *23*, 13–26. [CrossRef]

39. Kojok, F.; Fardoun, F.; Younes, R.; Outbib, R. Hybrid cooling systems: A review and an optimized selection scheme. *Renew. Sustain. Energy Rev.* **2016**, *65*, 57–80. [CrossRef]

40. Bhamare, D.K.; Rathod, M.K.; Banerjee, J. Passive cooling techniques for building and their applicability in different climatic zones-The state of art. *Energy Build.* **2019**, *198*, 467–490. [CrossRef]

41. Jiang, Y.; Lu, Z.; Wang, Z.; Lin, B. Review of thermal comfort infused with the latest big data and modeling progresses in public health. *Build. Environ.* **2019**, *164*. [CrossRef]

42. Moral-Munoz, J.A.; López-Herrera, A.G.; Herrera-Viedma, E.; Cobo, M.J. Science Mapping Analysis Software Tools: A Review. In *Handbook of Science and Technology Indicators*; Glänzel, W., Moed, H.F., Schmoch, U., Thelwall, M., Eds.; Springer International Publishing: Cham, Switzerland, 2019; ISBN 978-3-030-02511-3.

43. Small, H. Co-citation in the scientific literature: A new measure of the relationship between two documents. *J. Am. Soc. Inf. Sci.* **1973**, *24*, 265–269. [CrossRef]

44. Kessler, M.M. Bibliographic coupling between scientific papers. *Am. Doc.* **1963**, *14*, 10–25. [CrossRef]

45. Small, H.G.; Koenig, M.E.D. Journal clustering using a bibliographic coupling method. *Inf. Process. Manag.* **1977**, *13*, 277–288. [CrossRef]

46. Jarneving, B. Bibliographic coupling and its application to research-front and other core documents. *J. Informetr.* **2007**, *1*, 287–307. [CrossRef]

47. Bastian, M.; Heymann, S.; Jacomy, M. Gephi: An open source software for exploring and manipulating networks. In Proceedings of the Third International AAAI Conference on Weblogs and Social Media, San Jose, CA, USA, 17–20 May 2009.

48. Wang, Y.; Lai, N.; Zuo, J.; Chen, G.; Du, H. Characteristics and trends of research on waste-to-energy incineration: A bibliometric analysis, 1999–2015. *Renew. Sustain. Energy Rev.* **2016**, *66*, 95–104. [CrossRef]

49. Bartolini, M.; Bottani, E.; Grosse, E.H. Green warehousing: Systematic literature review and bibliometric analysis. *J. Clean. Prod.* **2019**, *226*, 242–258. [CrossRef]

50. Leydesdorff, L.; Rafols, I. Interactive overlays: A new method for generating global journal maps from Web-of-Science data. *J. Informetr.* **2012**, *6*, 318–332. [CrossRef]

51. Blondel, V.D.; Guillaume, J.-L.; Lambiotte, R.; Lefebvre, E. Fast unfolding of communities in large networks. *J. Stat. Mech. Theory Exp.* **2008**, *2008*, P10008. [CrossRef]

52. Lancichinetti, A.; Fortunato, S. Community detection algorithms: A comparative analysis. *Phys. Rev. E* **2009**, *80*, 56117. [CrossRef] [PubMed]

53. Rosenthal, D.H.; Gruenspecht, H.K.; Moran, E.A. Effects of Global Warming on Energy use for space heating and cooling in the United-States. *Energy J.* **1995**, *16*, 77–96. [CrossRef]

54. Balaras, C.A.; Droutsa, K.; Argiriou, A.A.; Asimakopoulos, D.N. Potential for energy conservation in apartment buildings. *Energy Build.* **2000**, *31*, 143–154. [CrossRef]

55. de Dear, R.J.; Brager, G.S. Thermal comfort in naturally ventilated buildings: Revisions to ASHRAE Standard 55. *Energy Build.* **2002**, *34*, 549–561. [CrossRef]

56. Kemp, P.C.; Neumeister-Kemp, H.G.; Esposito, B.; Lysek, G.; Murray, F. Changes in airborne fungi from the outdoors to indoor air; Large HVAC systems in nonproblem buildings in two different climates. *AIHA J.* **2003**, *64*, 269–275. [CrossRef] [PubMed]

57. Tsai, F.C.; Macher, J.M. Concentrations of airborne culturable bacteria in 100 large US office buildings from the BASE study. *Indoor Air* **2005**, *15*, 71–81. [CrossRef] [PubMed]

58. Akbari, H.; Levinson, R.; Rainer, L. Monitoring the energy-use effects of cool roofs on California commercial buildings. *Energy Build.* **2005**, *37*, 1007–1016. [CrossRef]

59. Levinson, R.; Akbari, H.; Konopacki, S.; Bretz, S. Inclusion of cool roofs in nonresidential Title 24 prescriptive requirements. *Energy Policy* **2005**, *33*, 151–170. [CrossRef]

60. van Hoof, J. Forty years of Fanger's model of thermal comfort: Comfort for all? *Indoor Air* **2008**, *18*, 182–201. [CrossRef] [PubMed]

61. Yang, L.; Yan, H.; Lam, J.C. Thermal comfort and building energy consumption implications—A review. *Appl. Energy* **2014**, *115*, 164–173. [CrossRef]

62. Ma, M.; Cai, W. What drives the carbon mitigation in Chinese commercial building sector? Evidence from decomposing an extended Kaya identity. *Sci. Total Environ.* **2018**, *634*, 884–899. [CrossRef]

63. Balaras, C.A.; Grossman, G.; Henning, H.-M.; Ferreira, C.A.; Podesser, E.; Wang, L.; Wiemken, E. Solar air conditioning in Europe—An overview. *Renew. Sustain. Energy Rev.* **2007**, *11*, 299–314. [CrossRef]

64. de Dear, R.; Brager, G.S. The adaptive model of thermal comfort and energy conservation in the built environment. *Int. J. Biometeorol.* **2001**, *45*, 100–108. [CrossRef] [PubMed]

65. Li, D.H.W.; Yang, L.; Lam, J.C. Zero energy buildings and sustainable development implications—A review. *Energy* **2013**, *54*, 1–10. [CrossRef]

66. Bolatturk, A. Determination of optimum insulation thickness for building walls with respect to various fuels and climate zones in Turkey. *Appl. Therm. Eng.* **2006**, *26*, 1301–1309. [CrossRef]

67. Li, D.H.W.; Yang, L.; Lam, J.C. Impact of climate change on energy use in the built environment in different climate zones—A review. *Energy* **2012**, *42*, 103–112. [CrossRef]

68. Yang, L.; Lam, J.C.; Tsang, C.L. Energy performance of building envelopes in different climate zones in China. *Appl. Energy* **2008**, *85*, 800–817. [CrossRef]

69. Hoyt, T.; Arens, E.; Zhang, H. Extending air temperature setpoints: Simulated energy savings and design considerations for new and retrofit buildings. *Build. Environ.* **2015**, *88*, 89–96. [CrossRef]

70. Jiang-Jiang, W.; Chun-Fa, Z.; You-Yin, J. Multi-criteria analysis of combined cooling, heating and power systems in different climate zones in China. *Appl. Energy* **2010**, *87*, 1247–1259. [CrossRef]

71. Hong, T.; Piette, M.A.; Chen, Y.; Lee, S.H.; Taylor-Lange, S.C.; Zhang, R.; Sun, K.; Price, P. Commercial Building Energy Saver: An energy retrofit analysis toolkit. *Appl. Energy* **2015**, *159*, 298–309. [CrossRef]

72. Mago, P.J.; Hueffed, A.; Chamra, L.M. Analysis and optimization of the use of CHP-ORC systems for small commercial buildings. *Energy Build.* **2010**, *42*, 1491–1498. [CrossRef]

73. Ucar, A.; Balo, F. Effect of fuel type on the optimum thickness of selected insulation materials for the four different climatic regions of Turkey. *Appl. Energy* **2009**, *86*, 730–736. [CrossRef]

74. Lam, J.C.; Wan, K.K.W.; Tsang, C.L.; Yang, L. Building energy efficiency in different climates. *Energy Convers. Manag.* **2008**, *49*, 2354–2366. [CrossRef]

75. de Dear, R.; Brager, G.S.; Cooper, D. Developing an adaptive model of thermal comfort and preference. *ASHRAE Trans.* **1998**, *104*, 1–18.

76. Brager, G.S.; de Dear, R.J. Thermal adaptation in the built environment: A literature review. *Energy Build.* **1998**, *27*, 83–96. [CrossRef]

77. Wan, K.K.W.; Li, D.H.W.; Liu, D.; Lam, J.C. Future trends of building heating and cooling loads and energy consumption in different climates. *Build. Environ.* **2011**, *46*, 223–234. [CrossRef]

78. Fanger, P.O. *Thermal Comfort. Analysis and Applications in Environmental Engineering*; Danish Technical Press: Copenhagen, Denmark, 1970.

79. Ole Fanger, P.; Toftum, J. Extension of the PMV model to non-air-conditioned buildings in warm climates. *Energy Build.* **2002**, *34*, 533–536. [CrossRef]

80. Humphreys, M.A.; Nicol, J.F. Understanding the adaptive approach to thermal comfort. In Proceedings of the 1998 ASHRAE Winter Meeting, San Francisco, CA, USA, 17–21 January 1998; Oxford University: Oxford, UK; Volume 104, pp. 991–1004.

81. Yao, R.; Li, B.; Steemers, K. Energy policy and standard for built environment in China. *Renew. Energy* **2005**, *30*, 1973–1988. [CrossRef]

82. Chwieduk, D. Towards sustainable-energy buildings. *Appl. Energy* **2003**, *76*, 211–217. [CrossRef]

83. Mago, P.J.; Chamra, L.M. Analysis and optimization of CCHP systems based on energy, economical, and environmental considerations. *Energy Build.* **2009**, *41*, 1099–1106. [CrossRef]

84. Medrano, M.; Brouwer, J.; McDonell, V.; Mauzey, J.; Samuelsen, S. Integration of distributed generation systems into generic types of commercial buildings in California. *Energy Build.* **2008**, *40*, 537–548. [CrossRef]

85. Wan, K.K.W.; Li, D.H.W.; Pan, W.; Lam, J.C. Impact of climate change on building energy use in different climate zones and mitigation and adaptation implications. *Appl. Energy* **2012**, *97*, 274–282. [CrossRef]

86. Wang, J.; Zhai, Z.J.; Zhang, C.; Jing, Y. Environmental impact analysis of BCHP system in different climate zones in China. *Energy* **2010**, *35*, 4208–4216. [CrossRef]

87. Wang, J.; Zhai, Z.J.; Jing, Y.; Zhang, C. Influence analysis of building types and climate zones on energetic, economic and environmental performances of BCHP systems. *Appl. Energy* **2011**, *88*, 3097–3112. [CrossRef]

88. Ozkahraman, H.T.; Bolatturk, A. The use of tuff stone cladding in buildings for energy conservation. *Constr. Build. Mater.* **2006**, *20*, 435–440. [CrossRef]

89. Manzano-Agugliaro, F.; Montoya, F.G.; Sabio-Ortega, A.; Garcia-Cruz, A. Review of bioclimatic architecture strategies for achieving thermal comfort. *Renew. Sustain. Energy Rev.* **2015**, *49*, 736–755. [CrossRef]

90. Alaidroos, A.; Krarti, M. Optimal design of residential building envelope systems in the Kingdom of Saudi Arabia. *Energy Build.* **2015**, *86*, 104–117. [CrossRef]

91. Okeil, A. A holistic approach to energy efficient building forms. *Energy Build.* **2010**, *42*, 1437–1444. [CrossRef]

92. Scherba, A.; Sailor, D.J.; Rosenstiel, T.N.; Wamser, C.C. Modeling impacts of roof reflectivity, integrated photovoltaic panels and green roof systems on sensible heat flux into the urban environment. *Build. Environ.* **2011**, *46*, 2542–2551. [CrossRef]

93. Olivieri, F.; Di Perna, C.; D'Orazio, M.; Olivieri, L.; Neila, J. Experimental measurements and numerical model for the summer performance assessment of extensive green roofs in a Mediterranean coastal climate. *Energy Build.* **2013**, *63*, 1–14. [CrossRef]

94. Susorova, I.; Angulo, M.; Bahrami, P.; Stephens, B. A model of vegetated exterior facades for evaluation of wall thermal performance. *Build. Environ.* **2013**, *67*, 1–13. [CrossRef]

95. Roman, K.K.; O'Brien, T.; Alvey, J.B.; Woo, O. Simulating the effects of cool roof and PCM (phase change materials) based roof to mitigate UHI (urban heat island) in prominent US cities. *Energy* **2016**, *96*, 103–117. [CrossRef]

96. Alam, M.; Jamil, H.; Sanjayan, J.; Wilson, J. Energy saving potential of phase change materials in major Australian cities. *Energy Build.* **2014**, *78*, 192–201. [CrossRef]

97. Zwanzig, S.D.; Lian, Y.; Brehob, E.G. Numerical simulation of phase change material composite wallboard in a multi-layered building envelope. *Energy Convers. Manag.* **2013**, *69*, 27–40. [CrossRef]

98. Saffari, M.; de Gracia, A.; Fernandez, C.; Cabeza, L.F. Simulation-based optimization of PCM melting temperature to improve the energy performance in buildings. *Appl. Energy* **2017**, *202*, 420–434. [CrossRef]

99. Saffari, M.; de Gracia, A.; Ushak, S.; Cabeza, L.F. Economic impact of integrating PCM as passive system in buildings using Fanger comfort model. *Energy Build.* **2016**, *112*, 159–172. [CrossRef]

100. van Hove, L.W.A.; Jacobs, C.M.J.; Heusinkveld, B.G.; Elbers, J.A.; van Driel, B.L.; Holtslag, A.A.M. Temporal and spatial variability of urban heat island and thermal comfort within the Rotterdam agglomeration. *Build. Environ.* **2015**, *83*, 91–103. [CrossRef]

101. Yang, F.; Qian, F.; Lau, S.S.Y. Urban form and density as indicators for summertime outdoor ventilation potential: A case study on high-rise housing in Shanghai. *Build. Environ.* **2013**, *70*, 122–137. [CrossRef]

102. Morakinyo, T.E.; Dahanayake, K.K.C.; Ng, E.; Chow, C.L. Temperature and cooling demand reduction by green-roof types in different climates and urban densities: A co-simulation parametric study. *Energy Build.* **2017**, *145*, 226–237. [CrossRef]

103. Herter, K.; Wayland, S. Residential response to critical-peak pricing of electricity: California evidence. *Energy* **2010**, *35*, 1561–1567. [CrossRef]

104. Yin, R.; Xu, P.; Piette, M.A.; Kiliccote, S. Study on Auto-DR and pre-cooling of commercial buildings with thermal mass in California. *Energy Build.* **2010**, *42*, 967–975. [CrossRef]

105. Turner, W.J.N.; Walker, I.S.; Roux, J. Peak load reductions: Electric load shifting with mechanical pre-cooling of residential buildings with low thermal mass. *Energy* **2015**, *82*, 1057–1067. [CrossRef]

106. Bartlett, K.H.; Kennedy, S.M.; Brauer, M.; van Netten, C.; Dill, B. Evaluation and determinants of airborne bacterial concentrations in school classrooms. *J. Occup. Environ. Hyg.* **2004**, *1*, 639–647. [CrossRef] [PubMed]

107. Rintala, H.; Pitkaranta, M.; Taubel, M. Microbial Communities Associated with House Dust. In *Advances in Applied Microbiology*; Laskin, A.I., Sariaslani, S., Gadd, G., Eds.; Elsevier Inc.: Amsterdam, The Netherlands, 2012; Volume 78, ISBN 978-0-12-394805-2.

108. Cardona, E.; Piacentino, A.; Cardona, F. Energy saving in airports by trigeneration. Part I: Assessing economic and technical potential. *Appl. Therm. Eng.* **2006**, *26*, 1427–1436. [CrossRef]

109. Balaras, C.A.; Dascalaki, E.; Gaglia, A.; Droutsa, K. Energy conservation potential, HVAC installations and operational issues in Hellenic airports. *Energy Build.* **2003**, *35*, 1105–1120. [CrossRef]

110. Mago, P.J.; Chamra, L.M.; Hueffed, A. A review on energy, economical, and environmental benefits of the use of CHP systems for small commercial buildings for the North American climate. *Int. J. Energy Res.* **2009**, *33*, 1252–1265. [CrossRef]

111. Jing, Y.-Y.; Bai, H.; Wang, J.-J.; Liu, L. Life cycle assessment of a solar combined cooling heating and power system in different operation strategies. *Appl. Energy* **2012**, *92*, 843–853. [CrossRef]

112. Jing, Y.-Y.; Bai, H.; Wang, J.-J. Multi-objective optimization design and operation strategy analysis of BCHP system based on life cycle assessment. *Energy* **2012**, *37*, 405–416. [CrossRef]

113. Smith, A.D.; Mago, P.J. Effects of load-following operational methods on combined heat and power system efficiency. *Appl. Energy* **2014**, *115*, 337–351. [CrossRef]

114. Gu, Q.; Ren, H.; Gao, W.; Ren, J. Integrated assessment of combined cooling heating and power systems under different design and management options for residential buildings in Shanghai. *Energy Build.* **2012**, *51*, 143–152. [CrossRef]

115. Romero Rodriguez, L.; Salmeron Lissen, J.M.; Sanchez Ramos, J.; Rodriguez Jara, E.A.; Alvarez Dominguez, S. Analysis of the economic feasibility and reduction of a building's energy consumption and emissions when integrating hybrid solar thermal/PV/micro-CHP systems. *Appl. Energy* **2016**, *165*, 828–838. [CrossRef]

116. Ucar, A. The environmental impact of optimum insulation thickness for external walls and flat roofs of building in Turkey's different degree-day regions. *Energy Educ. Sci. Technol. Part A Energy Sci. Res.* **2009**, *24*, 49–69.

117. Ekici, B.B.; Gulten, A.A.; Aksoy, U.T. A study on the optimum insulation thicknesses of various types of external walls with respect to different materials, fuels and climate zones in Turkey. *Appl. Energy* **2012**, *92*, 211–217. [CrossRef]

118. Bouden, C.; Ghrab, N. An adaptive thermal comfort model for the Tunisian context: A field study results. *Energy Build.* **2005**, *37*, 952–963. [CrossRef]

119. Yang, W.; Zhang, G. Thermal comfort in naturally ventilated and air-conditioned buildings in humid subtropical climate zone in China. *Int. J. Biometeorol.* **2008**, *52*, 385–398. [CrossRef] [PubMed]

120. Candido, C.; de Dear, R.J.; Lamberts, R.; Bittencourt, L. Air movement acceptability limits and thermal comfort in Brazil's hot humid climate zone. *Build. Environ.* **2010**, *45*, 222–229. [CrossRef]

121. Yao, R.; Li, B.; Steemers, K.; Short, A. Assessing the natural ventilation cooling potential of office buildings in different climate zones in China. *Renew. Energy* **2009**, *34*, 2697–2705. [CrossRef]

122. van Hoof, J.; Hensen, J.L.M. Quantifying the relevance of adaptive thermal comfort models in moderate thermal climate zones. *Build. Environ.* **2007**, *42*, 156–170. [CrossRef]

123. Chow, T.T.; Fong, K.F.; Givoni, B.; Lin, Z.; Chan, A.L.S. Thermal sensation of Hong Kong people with increased air speed, temperature and humidity in air-conditioned environment. *Build. Environ.* **2010**, *45*, 2177–2183. [CrossRef]

124. Singh, M.K.; Mahapatra, S.; Atreya, S.K. Adaptive thermal comfort model for different climatic zones of North-East India. *Appl. Energy* **2011**, *88*, 2420–2428. [CrossRef]

125. Mishra, A.K.; Ramgopal, M. Field studies on human thermal comfort—An overview. *Build. Environ.* **2013**, *64*, 94–106. [CrossRef]

126. Manu, S.; Shukla, Y.; Rawal, R.; Thomas, L.E.; de Dear, R. Field studies of thermal comfort across multiple climate zones for the subcontinent: India Model for Adaptive Comfort (IMAC). *Build. Environ.* **2016**, *98*, 55–70. [CrossRef]

127. de Dear, R.; Kim, J.; Candido, C.; Deuble, M. Adaptive thermal comfort in Australian school classrooms. *Build. Res. Inf.* **2015**, *43*, 383–398. [CrossRef]

128. Zomorodian, Z.S.; Tahsildoost, M.; Hafezi, M. Thermal comfort in educational buildings: A review article. *Renew. Sustain. Energy Rev.* **2016**, *59*, 895–906. [CrossRef]

129. Moon, J.W.; Han, S.-H. Thermostat strategies impact on energy consumption in residential buildings. *Energy Build.* **2011**, *43*, 338–346. [CrossRef]
130. Bansal, V.; Mishra, R.; Das Agarwal, G.; Mathur, J. Performance analysis of integrated earth-air-tunnel-evaporative cooling system in hot and dry climate. *Energy Build.* **2012**, *47*, 525–532. [CrossRef]
131. Theodosiou, T.G.; Ordoumpozanis, K.T. Energy, comfort and indoor air quality in nursery and elementary school buildings in the cold climatic zone of Greece. *Energy Build.* **2008**, *40*, 2207–2214. [CrossRef]
132. Dimoudi, A.; Kostarela, P. Energy monitoring and conservation potential in school buildings in the C′ climatic zone of Greece. *Renew. Energy* **2009**, *34*, 289–296. [CrossRef]
133. Perez, Y.V.; Capeluto, I.G. Climatic considerations in school building design in the hot-humid climate for reducing energy consumption. *Appl. Energy* **2009**, *86*, 340–348. [CrossRef]
134. Calise, F. Thermoeconomic analysis and optimization of high efficiency solar heating and cooling systems for different Italian school buildings and climates. *Energy Build.* **2010**, *42*, 992–1003. [CrossRef]
135. Dascalaki, E.G.; Sermpetzoglou, V.G. Energy performance and indoor environmental quality in Hellenic schools. *Energy Build.* **2011**, *43*, 718–727. [CrossRef]
136. Jokisalo, J.; Kurnitski, J.; Korpi, M.; Kalamees, T.; Vinha, J. Building leakage, infiltration, and energy performance analyses for Finnish detached houses. *Build. Environ.* **2009**, *44*, 377–387. [CrossRef]
137. Urbikain, M.K.; Sala, J.M. Analysis of different models to estimate energy savings related to windows in residential buildings. *Energy Build.* **2009**, *41*, 687–695. [CrossRef]
138. Chan, W.R.; Joh, J.; Sherman, M.H. Analysis of air leakage measurements of US houses. *Energy Build.* **2013**, *66*, 616–625. [CrossRef]
139. Didone, E.L.; Wagner, A. Semi-transparent PV windows: A study for office buildings in Brazil. *Energy Build.* **2013**, *67*, 136–142. [CrossRef]
140. Peng, J.; Curcija, D.C.; Lu, L.; Selkowitz, S.E.; Yang, H.; Zhang, W. Numerical investigation of the energy saving potential of a semi-transparent photovoltaic double-skin facade in a cool-summer Mediterranean climate. *Appl. Energy* **2016**, *165*, 345–356. [CrossRef]
141. Hygh, J.S.; de Carolis, J.F.; Hill, D.B.; Ranjithan, S.R. Multivariate regression as an energy assessment tool in early building design. *Build. Environ.* **2012**, *57*, 165–175. [CrossRef]
142. de Rosa, M.; Bianco, V.; Scarpa, F.; Tagliafico, L.A. Heating and cooling building energy demand evaluation; a simplified model and a modified degree days approach. *Appl. Energy* **2014**, *128*, 217–229. [CrossRef]
143. Shen, E.; Hu, J.; Patel, M. Energy and visual comfort analysis of lighting and daylight control strategies. *Build. Environ.* **2014**, *78*, 155–170. [CrossRef]
144. Asdrubali, F.; Bonaut, M.; Battisti, M.; Venegas, M. Comparative study of energy regulations for buildings in Italy and Spain. *Energy Build.* **2008**, *40*, 1805–1815. [CrossRef]
145. Eskin, N.; Tuerkmen, H. Analysis of annual heating and cooling energy requirements for office buildings in different climates in Turkey. *Energy Build.* **2008**, *40*, 763–773. [CrossRef]
146. Jaber, S.; Ajib, S. Thermal and economic windows design for different climate zones. *Energy Build.* **2011**, *43*, 3208–3215. [CrossRef]
147. Kim, K.-H. A comparative life cycle assessment of a transparent composite facade system and a glass curtain wall system. *Energy Build.* **2011**, *43*, 3436–3445. [CrossRef]
148. Ihm, P.; Krarti, M. Design optimization of energy efficient residential buildings in Tunisia. *Build. Environ.* **2012**, *58*, 81–90. [CrossRef]
149. Susorova, I.; Tabibzadeh, M.; Rahman, A.; Clack, H.L.; Elnimeiri, M. The effect of geometry factors on fenestration energy performance and energy savings in office buildings. *Energy Build.* **2013**, *57*, 6–13. [CrossRef]
150. Roetzel, A.; Tsangrassoulis, A.; Dietrich, U. Impact of building design and occupancy on office comfort and energy performance in different climates. *Build. Environ.* **2014**, *71*, 165–175. [CrossRef]
151. Li, C.; Hong, T.; Yan, D. An insight into actual energy use and its drivers in high-performance buildings. *Appl. Energy* **2014**, *131*, 394–410. [CrossRef]
152. Delgarm, N.; Sajadi, B.; Delgarm, S.; Kowsary, F. A novel approach for the simulation-based optimization of the buildings energy consumption using NSGA-II: Case study in Iran. *Energy Build.* **2016**, *127*, 552–560. [CrossRef]
153. Konis, K.; Gamas, A.; Kensek, K. Passive performance and building form: An optimization framework for early-stage design support. *Sol. Energy* **2016**, *125*, 161–179. [CrossRef]
154. El Fouih, Y.; Stabat, P.; Riviere, P.; Hoang, P.; Archambault, V. Adequacy of air-to-air heat recovery ventilation system applied in low energy buildings. *Energy Build.* **2012**, *54*, 29–39. [CrossRef]
155. Ke, Z.; Yanming, K. Applicability of air-to-air heat recovery ventilators in China. *Appl. Therm. Eng.* **2009**, *29*, 830–840. [CrossRef]
156. Liu, J.; Li, W.; Liu, J.; Wang, B. Efficiency of energy recovery ventilator with various weathers and its energy saving performance in a residential apartment. *Energy Build.* **2010**, *42*, 43–49. [CrossRef]
157. Ren, X.; Yan, D.; Wang, C. Air-conditioning usage conditional probability model for residential buildings. *Build. Environ.* **2014**, *81*, 172–182. [CrossRef]
158. Zhao, J.; Lasternas, B.; Lam, K.P.; Yun, R.; Loftness, V. Occupant behavior and schedule modeling for building energy simulation through office appliance power consumption data mining. *Energy Build.* **2014**, *82*, 341–355. [CrossRef]

159. Ghahramani, A.; Zhang, K.; Dutta, K.; Yang, Z.; Becerik-Gerber, B. Energy savings from temperature setpoints and deadband: Quantifying the influence of building and system properties on savings. *Appl. Energy* **2016**, *165*, 930–942. [CrossRef]

160. da Graca, G.C.; Augusto, A.; Lerer, M.M. Solar powered net zero energy houses for southern Europe: Feasibility study. *Sol. Energy* **2012**, *86*, 634–646. [CrossRef]

161. Ronghui, Q.; Lin, L.; Yu, H. Energy performance of solar-assisted liquid desiccant air-conditioning system for commercial building in main climate zones. *Energy Convers. Manag.* **2014**, *88*, 749–757. [CrossRef]

162. Liu, Z.; Xu, W.; Qian, C.; Chen, X.; Jin, G. Investigation on the feasibility and performance of ground source heat pump (GSHP) in three cities in cold climate zone, China. *Renew. Energy* **2015**, *84*, 89–96. [CrossRef]

163. Liu, Z.; Xu, W.; Zhai, X.; Qian, C.; Chen, X. Feasibility and performance study of the hybrid ground-source heat pump system for one office building in Chinese heating dominated areas. *Renew. Energy* **2017**, *101*, 1131–1140. [CrossRef]

164. Yang, L.; Lam, J.C.; Liu, J.; Tsang, C.L. Building energy simulation using multi-years and typical meteorological years in different climates. *Energy Convers. Manag.* **2008**, *49*, 113–124. [CrossRef]

165. Yang, L.; Wan, K.K.W.; Li, D.H.W.; Lam, J.C. A new method to develop typical weather years in different climates for building energy use studies. *Energy* **2011**, *36*, 6121–6129. [CrossRef]

166. Hong, T.; Chang, W.-K.; Lin, H.-W. A fresh look at weather impact on peak electricity demand and energy use of buildings using 30-year actual weather data. *Appl. Energy* **2013**, *111*, 333–350. [CrossRef]

167. Kalamees, T.; Jylha, K.; Tietavainen, H.; Jokisalo, J.; Ilomets, S.; Hyvonen, R.; Saku, S. Development of weighting factors for climate variables for selecting the energy reference year according to the EN ISO 15927-4 standard. *Energy Build.* **2012**, *47*, 53–60. [CrossRef]

168. Ren, Z.; Chen, Z.; Wang, X. Climate change adaptation pathways for Australian residential buildings. *Build. Environ.* **2011**, *46*, 2398–2412. [CrossRef]

169. Asimakopoulos, D.A.; Santamouris, M.; Farrou, I.; Laskari, M.; Saliari, M.; Zanis, G.; Giannakidis, G.; Tigas, K.; Kapsomenakis, J.; Douvis, C.; et al. Modelling the energy demand projection of the building sector in Greece in the 21st century. *Energy Build.* **2012**, *49*, 488–498. [CrossRef]

170. Yang, Y.; Li, B.; Yao, R. A method of identifying and weighting indicators of energy efficiency assessment in Chinese residential buildings. *Energy Policy* **2010**, *38*, 7687–7697. [CrossRef]

171. de la Flor, F.J.S.; Domínguez, S.Á.; Félix, J.L.M.; Falcón, R.G. Climatic zoning and its application to Spanish building energy performance regulations. *Energy Build.* **2008**, *40*, 1984–1990. [CrossRef]

172. Singh, M.K.; Mahapatra, S.; Atreya, S.K. Development of bio-climatic zones in north-east India. *Energy Build.* **2007**, *39*, 1250–1257. [CrossRef]

173. Shi, J.; Yang, L. A Climate Classification of China through k-Nearest-Neighbor and Sparse Subspace Representation. *J. Clim.* **2020**, *33*, 243–262. [CrossRef]

174. Yang, L.; Lyu, K.; Li, H.; Liu, Y. Building climate zoning in China using supervised classification-based machine learning. *Build. Environ.* **2020**, *171*, 106663. [CrossRef]

175. Kotharkar, R.; Ghosh, A.; Kotharkar, V. Estimating summertime heat stress in a tropical Indian city using Local Climate Zone (LCZ) framework. *Urban Clim.* **2021**, *36*, 100784. [CrossRef]

176. Ascione, F.; Bianco, N.; Mauro, G.M.; Napolitano, D.F.; Vanoli, G.P. Building heating demand vs climate: Deep insights to achieve a novel heating stress index and climatic stress curves. *J. Clean. Prod.* **2021**, *126616*. [CrossRef]

177. Bienvenido-Huertas, D.; Sánchez-García, D.; Rubio-Bellido, C.; Pulido-Arcas, J.A. Analysing the inequitable energy framework for the implementation of nearly zero energy buildings (nZEB) in Spain. *J. Build. Eng.* **2021**, *35*, 102011. [CrossRef]

178. Ávila-Delgado, J.; Robador, M.D.; Barrera-Vera, J.A.; Marrero, M. Glazing selection procedure for office building retrofitting in the Mediterranean climate in Spain. *J. Build. Eng.* **2021**, *33*, 101448. [CrossRef]

179. Bai, L.; Yang, L.; Song, B.; Liu, N. A new approach to develop a climate classification for building energy efficiency addressing Chinese climate characteristics. *Energy* **2020**, *195*, 116982. [CrossRef]

180. Aydin, N.; Biyikoğlu, A. Determination of optimum insulation thickness by life cycle cost analysis for residential buildings in Turkey. *Sci. Technol. Built Environ.* **2021**, *27*, 2–13. [CrossRef]

181. Akan, A.E. Determination and Modeling of Optimum Insulation Thickness for Thermal Insulation of Buildings in All City Centers of Turkey. *Int. J. Thermophys.* **2021**, *42*, 49. [CrossRef]

182. Bienvenido-Huertas, D. Analysis of the Impact of the Use Profile of HVAC Systems Established by the Spanish Standard to Assess Residential Building Energy Performance. *Sustainability* **2020**, *12*, 7153. [CrossRef]

183. Shrestha, M.; Rijal, H.B.; Kayo, G.; Shukuya, M. A field investigation on adaptive thermal comfort in school buildings in the temperate climatic region of Nepal. *Build. Environ.* **2021**, *190*, 107523. [CrossRef]

184. Guevara, G.; Soriano, G.; Mino-Rodriguez, I. Thermal comfort in university classrooms: An experimental study in the tropics. *Build. Environ.* **2021**, *187*, 107430. [CrossRef]

185. O'Neill, B.C.; Kriegler, E.; Riahi, K.; Ebi, K.L.; Hallegatte, S.; Carter, T.R.; Mathur, R.; van Vuuren, D.P. A new scenario framework for climate change research: The concept of shared socioeconomic pathways. *Clim. Chang.* **2014**, *122*, 387–400. [CrossRef]

186. Eyring, V.; Bony, S.; Meehl, G.A.; Senior, C.A.; Stevens, B.; Stouffer, R.J.; Taylor, K.E. Overview of the Coupled Model Intercomparison Project Phase 6 (CMIP6) experimental design and organization. *Geosci. Model Dev.* **2016**, *9*, 1937–1958. [CrossRef]

187. Edelenbosch, O.Y.; van Vuuren, D.P.; Blok, K.; Calvin, K.; Fujimori, S. Mitigating energy demand sector emissions: The integrated modelling perspective. *Appl. Energy* **2020**, *261*, 114347. [CrossRef]

188. Zheng, S.; Huang, G.; Zhou, X.; Zhu, X. Climate-change impacts on electricity demands at a metropolitan scale: A case study of Guangzhou, China. *Appl. Energy* **2020**, *261*, 114295. [CrossRef]

Article

Dynamics of Changes in Climate Zones and Building Energy Demand. A Case Study in Spain

Carmen Díaz-López [1], Joaquín Jódar [2], Konstantin Verichev [3], Miguel Luis Rodríguez [4], Manuel Carpio [5] and Montserrat Zamorano [1,*]

1 Department of Civil Engineering, ETS Ingenieria de Caminos, Canales y Puertos, Campus Fuentenueva s/n, University of Granada, 18071 Granada, Spain; carmendiaz@ugr.es
2 Department of Mathematics, Campus Las Lagunillas, University of Jaén, 23071 Jaén, Spain; jjodar@ujaen.es
3 Institute of Civil Engineering, Universidad Austral de Chile, General Lagos 2050, Valdivia 5111187, Chile; konstantin.verichev@uach.cl
4 Department of Applied Mathematics, Campus Fuentenueva, University of Granada, 18071 Granada, Spain; miguelrg@ugr.es
5 Department of Construction Engineering and Management, School of Engineering, Pontificia Universidad Católica de Chile, Avenida Vicuña Mackenna 4860, Santiago 7820436, Chile; manuel.carpio@ing.puc.cl
* Correspondence: zamorano@ugr.es; Tel.: +34-958-249-458

Abstract: In the current context of the climate crisis, it is essential to design buildings that can cope with climate dynamics throughout their life cycle. It will ensure the development of sustainable and resilient building stock. Thus, this study's primary objective has been to demonstrate that the current climatic zones for buildings in peninsular Spain do not represent the current climatic reality and are not adapted to climate change and the impact on the energy demand of buildings. For this reason, the climatic zones of 7967 peninsular cities have been updated and adapted to the RCP 4.5 and RCP 8.5 scenarios by using the data measured in 77 meteorological reference stations. The results obtained have shown that in more than 80% of the cities, buildings are designed and constructed according to an obsolete climatic classification that does not take into account the current or future climatic reality, which will significantly affect the thermal performance of a building and highlights the need to review the climatic zoning in the country. The results obtained can be extrapolated to other regions. The methodology defined in this work can be used as a reference, thus making an essential scientific contribution in reflecting on current capacities and the possibilities of improving the building stock.

Keywords: climate zone; climate change; building; energy demand; building resilience

Citation: Díaz-López, C.; Jódar, J.; Verichev, K.; Rodríguez, M.L.; Carpio, M.; Zamorano, M. Dynamics of Changes in Climate Zones and Building Energy Demand. A Case Study in Spain. *Appl. Sci.* **2021**, *11*, 4261. https://doi.org/10.3390/app11094261

Academic Editor: Luisa F. Cabeza

Received: 16 April 2021
Accepted: 5 May 2021
Published: 8 May 2021

1. Introduction

According to the fifth assessment report of the Intergovernmental Panel on Climate Change (IPCC), compared to values from 1850 to 1900, the global land surface temperature increased by 0.85 °C between 1880–2012 and 0.78 °C between 2003–2012 [1]. Projections of future climate conditions [1], predict an increase in global average temperature by the year 2100 in the range 1.4 to 5.8 °C [2], revealing an increasing disparity between historical climate patterns and current and future climate conditions resulting from anthropogenic changes [3].

This report also presents a set of four possible scenarios of climate change, called Representative Concentration Pathways (RCPs) [4]. These scenarios are characterized by the approximate calculation that gives the Radiative Forcing (RF) in the year 2100, with respect to the year 1750, taking into consideration different trajectories for emissions of long-lived greenhouse gases (LLGHGs) and short-lived air pollutants, the corresponding concentration levels and land use [3]. In the following, a description is given of the two scenarios within which the present study is realized. These scenarios were selected due to their wide application in other studies related to climate change, building EC, and climate zones for buildings [4].

RCP 4.5. An intermediate stabilization pathway in which RF is stabilized at approximately 4.5 W/m^2 after 2100. It will be necessary to limit emissions through increased use of electricity, lower-emission energies, CO_2 capture technologies, and geological storage. The area of forests is also expected to increase for this scenario, compared to the current state. Furthermore, CO_2 emissions from energy and industrial sources are expected to increase until 2040 and then decrease to the prescribed atmospheric CO_2 concentration of 538 ppm in 2100. At the same time, by 2081–2100, the global mean surface temperature will increase by 1.8 °C (likely range 1.1 to 2.6 °C) compared to the 1986–2005 climate period [5,6].

RCP 8.5. During the 21st century, RF will grow steadily and reach 8.5 W/m^2 in 2100. Very high GHG emissions characterize the scenario. RCP 8.5 combines the assumptions of a steady increase in the global population, a moderate rate of technological change, and energy intensity improvements. In the long term, this leads to high energy demand and GHG emissions in the absence of a climate change policy. The prescribed CO_2 concentration is 936 ppm in 2100. At the same time, by 2081–2100, the global mean surface temperature will increase by 3.7 °C (likely range 2.6 to 4.8 °C) compared to 1986–2005 [7,8].

Consequently, in the future, the temperature increase will determine the setting of new energy budgets; indeed, climate change is transforming buildings and cities' energy performance [8,9]. Thus, numerous studies show that it is essential to take climate dynamism into account in the building design phase; otherwise, the estimated energy demand (ED) could triple [9–12]. For example, Christenson et al. 2006 [13], showed how global warming's impact increases the cooling energy demand of buildings. In the study by de Rosa et al. 2014 [14], a simplified building dynamic model, based on the electrical analogy, was developed and implemented in a Matlab/Simulink environment in order to perform several analyses on heating (H) and cooling (C) energy demand in a wide range of climatic conditions. Verichev et al. 2020 [6] showed the effects of climate change on variations in climatic zones and heating energy consumption of residential buildings in southern Chile under different climate change scenarios. Thus, in the general context of global warming, in cold regions, the intensity of energy consumption reduction will be greater than the intensity of increase in cooling energy consumption [15,16]. Similarly, numerous studies conclude that in regions with a warm climate, the heating energy consumption (EC) of buildings will decrease. At the same time, the cooling EC will increase in conjunction with an increase in days of indoor thermal discomfort [17–25].

Currently, there are not many studies to assess the evolution of climatic zones for buildings. For example, in the study of Zhai and Helman 2019, the authors use data from 23 climate models, but the evolution of building climates is analyzed in only seven cities in the United States [26]. In China, the evolution of ASHRAE building climatic zones was analyzed based on data from 5 climate models for the RCP 4.5 scenario [27]. In the case of the analysis of the climatic zones of ASHRAE, it is quite convenient to use the data of climatic models on the daily maximum and minimum temperatures for future periods without additional modifications. On the other hand, some studies use the morphing methodologies described in the study of Belcher et al. 2005 [28]. For example, in the study of the dynamics of climatic zones for construction in Chile [6]; in study of risk of overheating increases of residential buildings in Sweden [29]; evaluation of life cycle impacts of buildings, integrating climate change effect and evolution of the energy mix in the long term in France [30].

At the same time, to assess thermal comfort or to simulate building energy consumption in future climatic conditions, it is necessary to use more complex methodologies for morphing meteorological data, taking into account changes in the intraday variability of meteorological parameters [31,32]. Additionally, in some studies, special tools are used to generate meteorological files for future periods, such as OZClim [33] or CCWorldWeatherGen [34].

In order to mitigate and adapt to these climate effects, the European Commission presented Directive (EU) 2018/844 [35] of the European Parliament and the Council of 30 May 2018 on energy efficiency, which, together with the European Green Deal [36], will adopt a new and more ambitious EU strategy on adaptation to climate change. In this context, many countries have developed regulations based on climate zone classification, a beneficial method to design buildings with lower energy consumption [37–39], and high thermal comfort [40]. This method is based on analyzing large amounts of meteorological, environmental, and social data to contribute to the search for climate models that absorb all of the above [41]. The number of climate zones depends on each country, the thresholds set, and the methodology used. For example, Spain, in the Technical Building Code (CTE in Spanish) [42], establishes 15 climate zones (α3, A2, A3, A4, B2, B3, B4, C1, C2, C3, C4, D1, D2, D3, E1) identified by a letter corresponding to the climatic severity in winter (α, A, B, C, D, and E) and a number (1, 2, 3, and 4) corresponding to the summer values. Portugal establishes nine climate zones (I1-V1, I1-V2, I1-V3, I2-V1, I2-V2, I2-V3, I3-V1, I3-V2, I3-V3) from the different combinations of winter letters (V1, V2, and V3) and summer letters (I1, I2, and I3) [43]; France establishes eight zones (H1a, H1b, H1c, H2a, H2b, H2c, H2d, H3), taking into account winter temperatures (H1, H2, and H3) and summer temperatures (a, b, c, and d) [44]. In any case, the climatic zonings used in different countries are based on the climatic series existing at the time of their formulation, and therefore do not allow the design of building parks capable of adapting to climatic dynamism [45].

Consequently, it is essential to design and construct buildings capable of assuming the climatic dynamics throughout their life cycle. Knowledge of the climatic reality will guarantee the development of a building stock that is certainly sustainable and resilient. For this reason, the main objective of this work has been to analyze the dynamics of changes in climate zones and their effect on the energy demand of buildings. Spain has been selected as the study area due to its climatic variety, which will allow the applied methodology, results, and conclusions obtained to be used as a reference in other regions. Furthermore, in this country, there is low investment in sustainable building, with the construction sector being one of its primary energy consumers, which translates into one of the highest consumption rates per Gross Domestic Product (GDP) in the European Union [46], highlighting the urgent need to take measures to solve this problem.

2. Materials and Methods

An update of the CTE climate zones [42] of 7967 localities of peninsular Spain has been carried out, under two of the four scenarios called Representative Concentration Pathways (RCPs) (RCP 2.6, RCP 4.5, RCP 6.0, and RCP 8.5), specifically RCP 4.5 and RCP 8.5 scenarios, to achieve the objective of this study. These scenarios are characterized by their approximate calculation of the total RF in the year 2100, relative to 1750 [6]. Each scenario describes a different trajectory for long-lived greenhouse gas emissions (LLGHGs) and short-lived air pollutants, the corresponding concentration levels, land use, and radiative forcing [5]. Besides, to know the effect of this dynamic of changes in buildings' energy consumption, a typical dwelling has been taken as a reference, and the evolution of its energy demand has been analyzed. The description of the bases for the definition of CTE climate zones and the methodology applied for this purpose is described below.

2.1. Basis for the Definition of CTE Climate Zones

In Spain, the CTE and its Basic Document on Energy Saving (DB-HE) [42] establishes a methodology that allows the definition of climatic zones for buildings. This methodology is based on the concept of climatic severity index (CSI), a unique number on a dimensionless scale that is specific for each geographical location [47] and that allows differentiating between climatic severity index for summer (SCS) and winter (WCS).

The WCS and SCS indices are obtained by applying Equations (1) and (2) [42], respectively, where $HDD20_{oct-may}$ is the sum of winter degree-days in 20 °C bases for the months ranging from October to May, calculated through the hourly method; $CDD20_{jun-sep}$ is the sum of summer degree-days in 20 °C bases for the months ranging from June to September, calculated through the hourly method; a, b, c, d, and e are the regression coefficients whose values are $a = 3.546 \times 10^{-4}$, $b = -4.043 \times 10^{-1}$, $c = 8.394 \times 10^{-8}$, $d = -7.325 \times 10^{-2}$, $e = -1.137 \times 10^{-1}$, in the case of WCS, and $a = 2.990 \times 10^{-3}$, $b = -1.1597 \times 10^{-7}$, $c = -1.713 \times 10^{-1}$, in the case of SCS; n is the sum of sunshine duration hours in the period from October to May; and N is the sum of the maximum possible of sunshine duration hours for the months from October to May.

$$WCS = a \cdot HDD20_{oct-may} + b \cdot \frac{n}{N} + c \cdot HDD20_{oct-may}{}^2 + d \cdot \left(\frac{n}{N}\right)^2 + e \qquad (1)$$

$$SCS = a \cdot CDD20_{jun-sep} + b \cdot CDD20_{jun-sep}{}^2 + c \qquad (2)$$

Each of the six winter climate zones defined in the DB-HE is assigned a letter (α, A, B, C, D, and E) corresponding to the WCS interval indicated in Table 1, with the climate zone α having the warmest winter and E the coldest [16]. By the four summer climate zones defined in the DB-HE and identified with a number (1, 2, 3, and 4), these are determined according to the SCS. Besides, it corresponds to the interval indicated in Table 1, being one the climate zone with the least warm summer and four the warmest [42]. Finally, the combination of letter and number given in Table 1 is the one that generates the building climate zone code for any city or geographical location. According to the provisions of the DB-HE document of CTE in peninsular Spain, there are 12 possible combinations and, as a result, climate zones (A3, A4, B3, B4, C1, C2, C3, C4, D1, D2, D3, and E1).

Table 1. Intervals for climate zoning.

Intervals for Winter Zoning					
α	A	B	C	D	E
WCS $\leq$ 0	0 < WCS $\leq$ 0.23	0.23 < WCS $\leq$ 0.5	0.5 < WCS $\leq$ 0.93	0.93 < WCS $\leq$ 1.51	WCS > 1.51
Intervals for Summer Zoning					
1	2	3	4		
SCS $\leq$ 0.5	0.5 < SCS $\leq$ 0.83	0.83 < SCS $\leq$ 1.38	SCS > 1.38		

2.2. Methodology

The methodology used to achieve the objectives of this work consists of four phases, which are described in the following sections:

(i)   Determination of climate severity indices.
(ii) Determination of the dynamics of changes in climate severity indices.
(iii) Proposal for updating climate zones for peninsular Spain.
(iv) Evaluation of the dynamics of changes in energy demand in housing.

2.2.1. Determination of Climate Severity Indices

From among the almost 800 weather stations located by the State Meteorological Agency (AEMET) in peninsular Spain [48], whose data can be provided for research, a total of 77 were selected (Figure 1 and Table 2). Considering their proximity to urban centers and a homogeneous distribution based on these centers' population, they were also sought with a minimum measurement period of 3 years, between 2015–2018, and which had hourly temperature measurement data available. As for the data relating to sunshine duration hours, 55 of them were able to provide them; in the case of the remaining 22 stations, the data from the geographically closest station that had them was used.

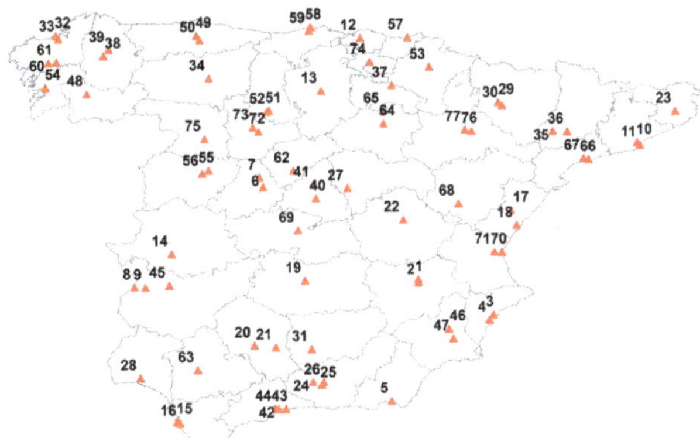

Figure 1. Geographical location of AEMET weather stations.

Table 2. Calculated indices of winter (WCS) and summer (SCS) climate severity for the period 2015–2018.

WS	Latitude	Longitude	Altitude	WCS	SCS	WS	Latitude	Longitude	Altitude	WCS	SCS
1	39.00556	−1.862222	676	0.94	1.64	40	40.45167	−3.724167	664	0.96	1.76
2	38.95417	−1.856389	702	0.92	1.53	41	40.69611	−3.765000	1004	1.11	1.19
3	38.37250	−0.494167	22	0.14	1.74	42	36.71667	−4.419722	25	−0.01	1.52
4	38.28278	−0.570833	43	0.23	1.82	43	36.71500	−4.286111	7	−0.06	1.91
5	36.84691	−2.356989	21	0.20	1.75	44	36.71778	−4.481667	54	0.04	1.90
6	40.65917	−4.680000	1130	1.44	0.88	45	38.91583	−6.385556	228	0.57	1.98
7	40.82889	−4.741944	920	1.36	1.04	46	38.12806	−1.305833	150	0.17	2.05
8	38.88750	−7.008056	175	0.56	2.08	47	37.95778	−1.228611	75	0.21	2.15
9	38.88333	−6.813889	185	0.49	2.08	48	42.32528	−7.859722	442	0.84	1.11
10	41.41833	2.124167	408	0.61	0.92	49	43.35333	−5.874167	336	0.90	0.12
11	41.37500	2.173889	5	0.23	1.48	50	43.27500	−5.819167	170	0.86	0.29
12	43.29806	−2.906389	42	0.74	0.43	51	42.00944	−4.560556	736	1.41	0.93
13	42.35694	−3.620278	891	1.61	0.57	52	41.99556	−4.602778	874	1.54	0.71
14	39.47139	−6.338889	10	0.62	2.04	53	42.77750	−1.649722	459	1.02	0.75
15	36.49972	−6.257778	2	−0.02	1.38	54	42.43833	−8.615833	108	0.69	0.43
16	36.46556	−6.205556	28	0.00	1.380	55	40.95750	−5.662222	775	1.15	1.12
17	40.22722	−0.169722	10	0.67	1.24	56	40.90361	−5.779722	817	1.45	1.00
18	39.95722	−0.071944	43	0.22	1.70	57	43.30639	−2.041111	251	0.82	0.09
19	38.98917	−3.920278	628	0.80	2.16	58	43.49111	−3.800556	52	0.57	0.04
20	37.84889	−4.846667	90	0.37	2.52	59	43.42861	−3.831389	3	0.65	0.30
21	37.81028	−4.462222	275	0.41	2.42	60	42.87611	−8.555833	240	0.83	0.37
22	40.06722	−2.131944	948	1.13	1.38	61	42.88806	−8.410556	370	0.97	0.32
23	41.97222	2.819444	76	0.69	1.45	62	40.94528	−4.126389	1005	1.28	1.04
24	37.18972	−3.789444	567	0.61	2.10	63	37.41667	−5.879167	34	0.16	2.420
25	37.13722	−3.631389	687	0.64	1.79	64	41.77500	−2.483056	1082	1.54	0.710
26	37.18972	−3.595556	775	0.60	1.89	65	41.99583	−2.494167	1260	1.87	0.360
27	40.63028	−3.150000	721	0.95	1.69	66	41.12389	1.249167	55	0.35	1.440
28	37.27833	−6.911667	19	0.17	1.74	67	41.14500	1.163611	71	0.44	1.270
29	42.13889	−0.395000	463	0.97	1.40	68	40.35056	−1.124167	900	1.28	1.090
30	42.08444	−0.325556	546	1.03	1.26	69	39.88472	−4.045278	515	0.74	2.200
31	37.77750	−3.808889	580	0.42	2.33	70	39.47972	−0.337500	6	0.22	1.430
32	43.36583	−8.421389	58	0.55	0.04	71	39.48500	−0.474722	56	0.34	1.650
33	43.30694	−8.371944	98	0.76	0.12	72	41.64083	−4.754444	735	1.27	1.120
34	42.58833	−5.651111	912	1.55	0.33	73	41.71194	−4.855556	846	1.51	0.760
35	41.62611	0.598056	185	0.89	1.63	74	42.87194	−2.732778	513	1.32	0.460
36	41.61694	0.866667	252	0.95	1.39	75	41.51556	−5.735278	656	1.18	1.260
37	42.45222	−2.331111	353	1.01	0.92	76	41.63333	−0.882222	258	0.34	1.580
38	42.99833	−7.552500	442	1.18	0.38	77	41.66056	−1.004167	249	0.72	1.550
39	43.11139	−7.457500	445	1.25	0.25						

At each of these stations, the WCS and SCS indices were calculated using as a basis for calculation Equations (1) and (2) of the CTE [42] described in the previous section. In the case of Equation (1), the values of N, for each geographical location of stations, were calculated with the application "NOAA solar calculations year" [49] by the NOAA Earth System Research Laboratory (ESRL) Global Monitoring Division.

2.2.2. Determining the Dynamics of Changes in Climate Severity Indices

The calculation of the changes for the WCS and SCS indices of each of the stations was carried out using the Climate Change Adaptation Platform (AdapteCCa) [50] which contains the results of daily minimum and maximum temperature projections for the RCP 4.5 and RCP 8.5 scenarios from a total of 16 climatological models.

Based on the projection data and current hourly temperature measurement data from AEMET, for each of the 77 meteorological stations, firstly, the difference values (or deltas) of the monthly average temperature between the baseline climate period (2018) and the two future periods (2055 and 2085) were calculated. Then, the hourly data from the AEMET stations were modified based on the monthly temperature deltas obtained. The modification of hourly temperature data has been carried out according to methodologies already presented in other scientific works [28,51,52], based on which it is possible to apply the "a shift" algorithm to modify baseline climate data, to modify hourly baseline climate temperature values by adding the projected monthly average difference for future years. For this purpose, Equations (3) and (4) [42] have been used to calculate HDD and CDD, respectively, in the future; where $HDD_{d,Y}$ and $CDD_{d,Y}$—are daily values of HDD and CDD in the future; $T_{i,2018}$—temperature of measurements in i-hour of the day in the year 2018; ΔT^j_{Y-2018}—delta of monthly temperatures in j-month between years in future (Y = 2055 and 2085) and baseline climate.

$$HDD_{d,Y} = \left[\sum_{i=1}^{24}\left(T_b - \left(T_{i,2018} + \Delta T^j_{Y-2018}\right)\right)^+\right]\frac{1}{24} \tag{3}$$

$$CDD_{d,Y} = \left[\sum_{i=1}^{24}\left(\left(T_{i,2018} + \Delta T^j_{Y-2018}\right) - T_b\right)^+\right]\frac{1}{24} \tag{4}$$

Based on the modified HDD and CDD results for the future, the WCS (Equation (1)) and SCS (Equation (2)) values for 2055 and 2085 were recalculated to account only for temperature changes without estimating changes in sunshine duration hours for the WCS index. This simplification was possible because the temperature in the climate models for the future already takes into account changes in atmospheric radiative conditions.

Finally, the WCS and SCS indices were calculated, and the dynamics of changes in the climate zones were obtained; for this purpose, the procedure followed for the calculation of the climate zone adaptation in the previous section was similar.

2.2.3. Proposed Update of Climate Zones for Peninsular Spain

The climatic zone classification for the 7967 Spanish municipalities of our research is based on determining their WCS and SCS indices.

Firstly, the computation of the WCS and SCS indices at the 77 weather stations as described above is carried out. It can be done by applying the formulae given because the values of temperature and sunshine duration hours required in the corresponding equations are available for those locations. These values are not available for the 7967 municipalities, and, as a consequence, their WCS and SCS indices cannot be calculated as done for the 77 weather stations. Their determination is then obtained by approximation, using an interpolation method based on radial basis functions (RBF).

For a given set of data (measurements and locations at which these measurements were obtained), the approximation procedure tries to determine a function ("approximating function") that is a good fit for the given data. In the approximation process using interpolation, this good fit is achieved by imposing that the approximating function's outputs exactly match the given measurements at the corresponding locations. Besides, information on the studied problem can also be deduced at locations different from where the measurements were obtained [53].

Approximation, and in particular interpolation employing RBF, has found significant applications in science, engineering, economics, biology, and medicine, among others. In our case, the determination of the WCS and SCS indices at the 7967 municipalities from the WCS and SCS indices calculated at the 77 weather stations was obtained by using an approximant expressed as a finite linear combination of a particular radial basis function and its translations (it is important to emphasize that the selected 77 weather stations are well distributed throughout peninsular Spain). To make this approximation, the inverse multiquadric function is given by the expression $\phi(r) = \frac{1}{\sqrt{1+(\varepsilon r)^2}}, r \geq 0$, was chosen as the basis function, but there are other possibilities. A wide range of radial basis functions can be found in the literature [54]. The parameter $\varepsilon \geq 0$ that appears in the above expression is a shape parameter.

Let us illustrate the determination of the climatic severity indices more precisely. For the case of the WCS index, an interpolant function $s\,(x, y, z)$ given by the expression:

$$s(x,y,z) = \sum_{i=1}^{77} a_i \phi(\| (x,y,z) - (x_i,y_i,z_i) \|)$$

is considered, where $(x_i, y_i, z_i) \in \mathrm{R}^3, i = 1, \ldots, 77$, represent the latitude, longitude, and altitude coordinates at each of the 77 weather stations, $\|.\|$ is the Euclidean norm on R^3, $\phi: [0, \infty) \to \mathrm{R}$ is the basis function, and $a_i, i = 1, \ldots, 77$, constitutes a set of real coefficients to be determined.

These coefficients are obtained by imposing that the output provided by the interpolant function s at each of the weather stations is the corresponding WCS index, which is known. Once the coefficients are calculated, the interpolant function s is therefore determined. The evaluation of s at any value $(x, y, z) \in \mathrm{R}^3$ corresponding to the latitude, longitude, and altitude coordinates at any peninsular Spanish location provides the searched approximation for the unknown WCS index at that location. The SCS case would be analogous.

This is the procedure followed for the 2015–2018 period. The corresponding one for the years 2055 and 2085 is utterly similar except that, for the starting stage, the WCS and SCS indices at the 77 weather stations, both for the RCP 4.5 and RCP 8.5 contexts, need to be recalculated, as described in Section 2.2.2.

Once the WCS and SCS indices are obtained at each of the 7967 municipalities, they can be classified inside the corresponding climatic zone. Remarkably, the main advantage of the interpolation method previously exhibited is that it provides a continuous function to compute the climatic severity indices at any location. Consequently, it could make possible a numerical climatic classification at any municipality instead of the more rigid one described by zones.

2.2.4. Assessment of the Dynamics of Changes in Energy Demand in Dwellings

Once the WCS and SCS indices have been calculated for the periods 2015–2018, 2055, and 2085, an analysis of the dynamics of changes in energy demand is carried out for the RCP 4.5 and RCP 8.5 scenarios in the 77 locations of the meteorological stations selected for the study. The city of Madrid has been taken as a geographical reference point, to which, according to the CTE, the values WCS = 1.0 and SCS = 1.0 [42] correspond; consequently, by multiplying the value of the energy demand of a dwelling located in Madrid by the value of the WCS (or SCS) index of any geographical location, it is possible to estimate the demand of that dwelling in that place.

For this analysis, an existing typical building of a six-storey multi-family block of flats, used in the work of López–Ochoa et al. 2017 [55] was considered. The block consists of a ground floor and five storeys. Its base measures 22 by 22 m, which is equivalent to an area of 484 m^2. The height of each floor is 3 m. The main facade faces north. Each of the five floors has 4 types of flats: Apartment A has 3 bedrooms and a size of 100.05 m^2; Apartment B has 3 bedrooms and a size of 101.93 m; Apartment C has 4 bedrooms and a size of 137.64 m^2; and Apartment D has 3 bedrooms and a size of 103.69 m^2.

The building's thermal transmittances are similar to the limit values set in CTE-DB-HE1 2009, fulfilling the requirements of CTE-DB-HE 2009 [56]. The heating and cooling energy demands are assessed using the official HULC tool [39]. In addition to determine these demands, this tool is used to verify compliance with the requirements of CTE-DB-HE1 2013. The energy demand for heating this house in Madrid is 42.74 kWh/m^2 year and for cooling is 14.09 kWh/m^2 year [55].

3. Results and Discussion

After applying the previous section's methodological steps, the results shown in Figures 2–7 and Tables 3–5 were obtained, which are analyzed and discussed below.

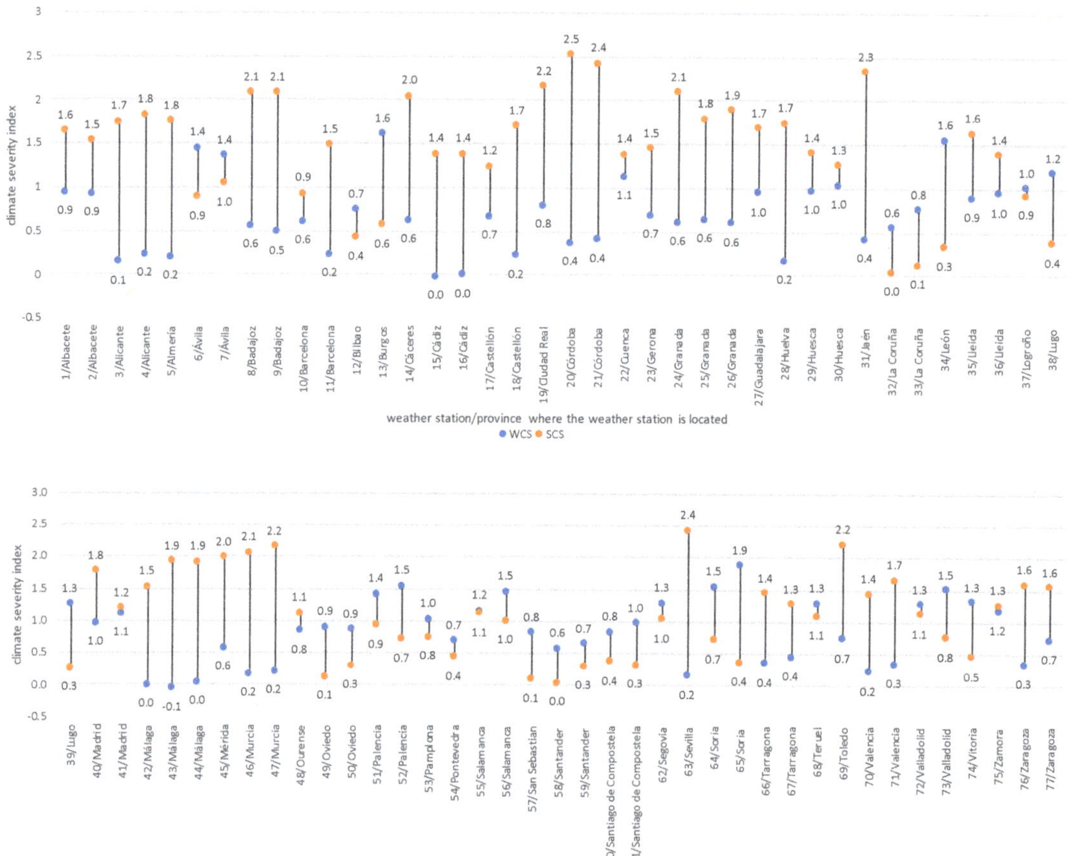

Figure 2. Calculated indices of winter (WCS) and summer (SCS) climate severity for the period 2015–2018.

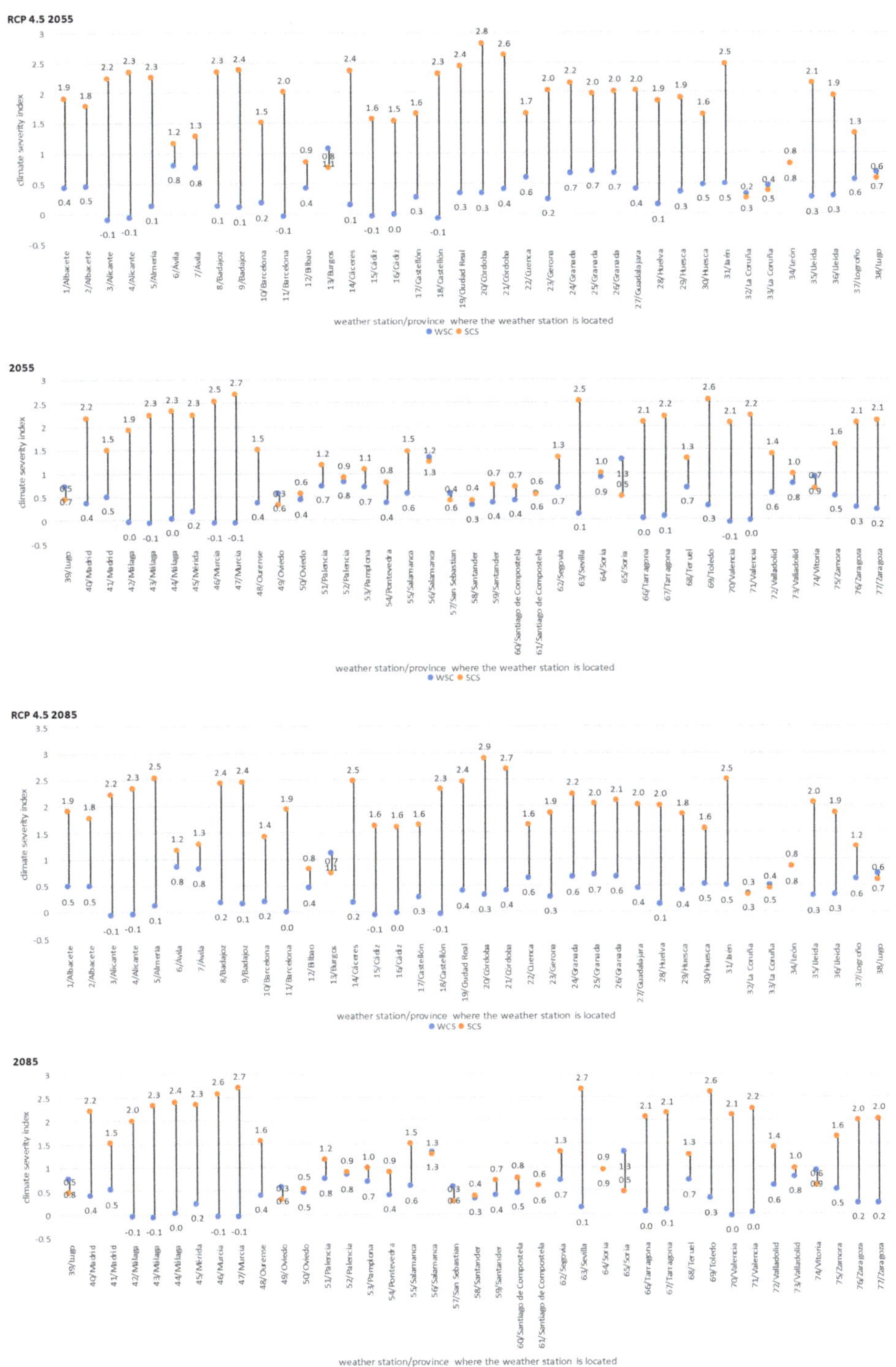

Figure 3. Calculated indices of winter (WCS) and summer (SCS) climate severity for the RCP 4.5.

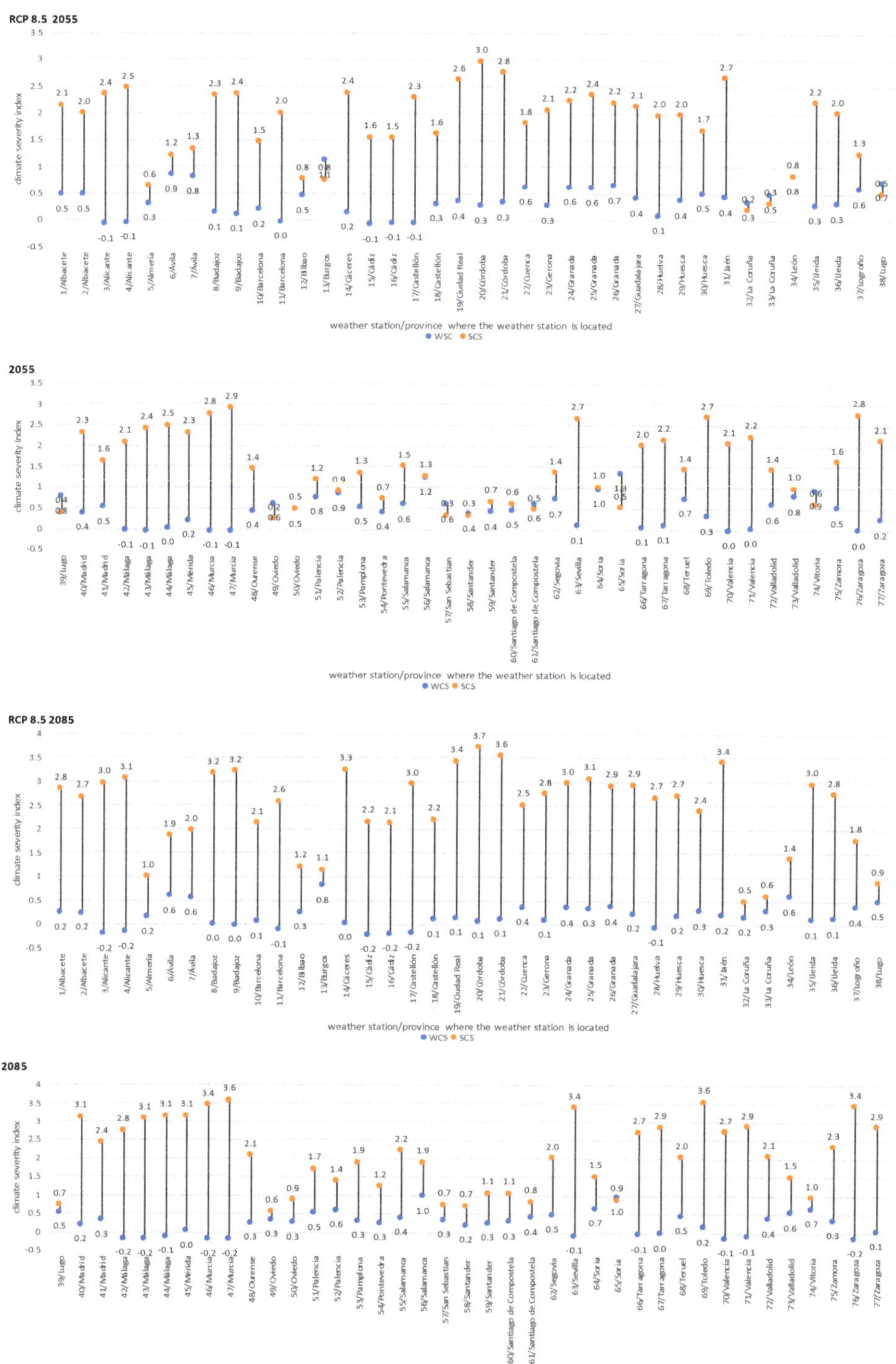

Figure 4. Calculated indices of winter (WCS) and summer (SCS) climate severity for the RCP 8.5.

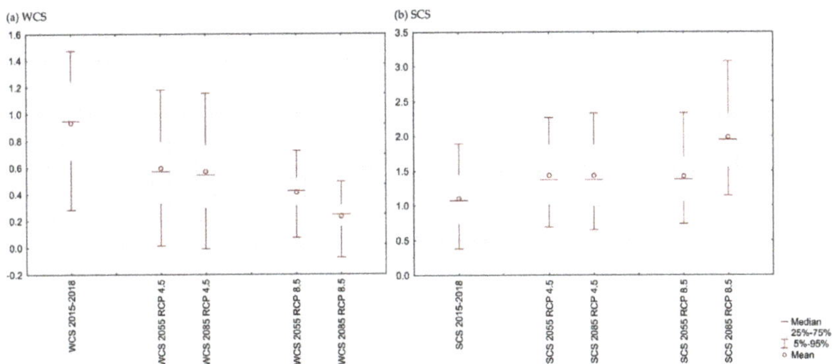

Figure 5. Boxplots of values of the WCS (**a**) and SCS (**b**) indexes in 7967 localities in mainland Spain for the period 2015–2018 and future periods.

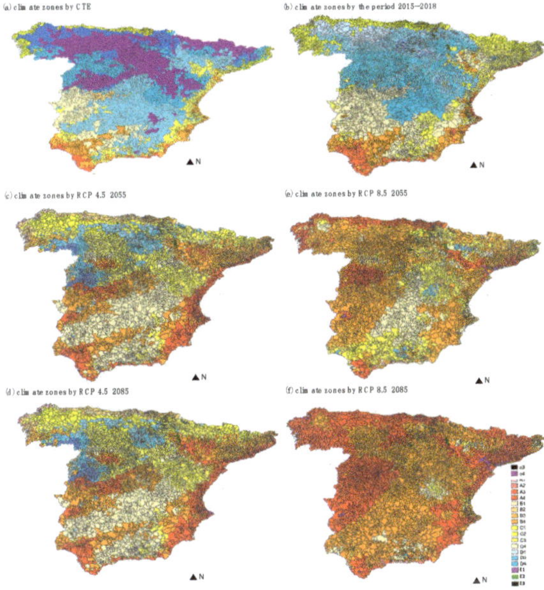

Figure 6. Maps of (**a**) CTE, (**b**) climatic zone by the period 2015–2018, (**c**) climate zones for RCP 4.5 2055, (**d**) climate zones for RCP 4.5 2085, (**e**) climate zones for RCP 8.5 2055, (**f**) climate zones for RCP 8.5 2085.

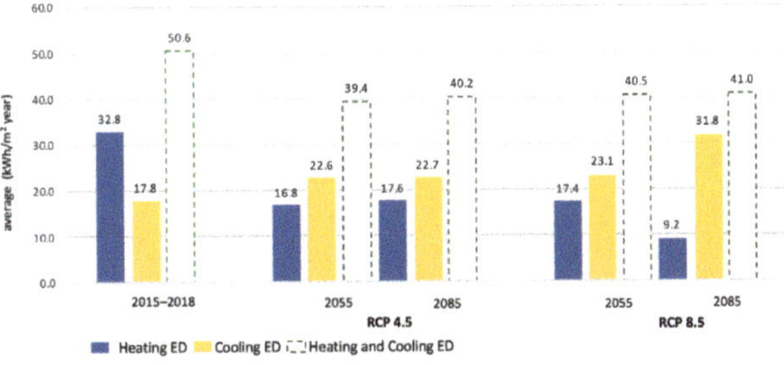

Figure 7. Dynamics of changes in the average of heating and cooling energy demand for the RCP 4.5 and RCP 8.5 scenarios by 77 stations.

Table 3. Dynamics of changes in climate zones according to scenarios.

Scenario	Winter Climatic Zone			Summer Climatic Zone			Winter + Summer Climatic Zone
	Code Number of CZ that Change	% of Cities Modifying CZ	% of Cities that Change CZ	Code Number of CZ that Change	% of Cities Modifying CZ	% of Cities that Change CZ	% of Cities that Change CZ
2015–2018	1	47.2%	52.1%	1	58.7%	72.0%	84%
	2	4.4%		2	12.2%		
	3	0.1%		3	0.2%		
	4	0.1%		4	0.0%		
	5	0.1%		−1	0.9%		
	−1	0.1%					
RCP 4.5 2055	1	44.8%	89.0%	1	47.7%	88.6%	98%
	2	40.5%		2	39.4%		
	3	3.0%		3	1.4%		
	4	0.0%		4	0.0%		
	5	0.0%		−1	0.1%		
	−1	0.7%					
RCP 4.5 2085	1	41.0%	89.9%	1	48.3%	87.6%	98%
	2	43.4%		2	38.1%		
	3	4.7%		3	1.2%		
	4	0.1%		4	0.0%		
	5	0.0%		−1	0.1%		
	−1	0.7%					
RCP 8.5 2055	1	24.4%	92.4%	1	33.0%	82.7%	97%
	2	33.1%		2	25.5%		
	3	26.5%		3	18.9%		
	4	4.1%		4	0.0%		
	5	0.0%		−1	4.1%		
	−1	3.8%					
RCP 8.5 2085	1	10.1%	100%	1	31.0%	91.0%	98%
	2	36.7%		2	26.3%		
	3	35.4%		3	32.2%		
	4	11.6%		4	0.0%		
	5	0.0%		−1	1.1%		
	−1	0.6%					

Table 4. Percentage of climate zones in different scenarios.

Climatic Zone	Present		RCP 4.5		RCP 8.5	
	CTE	2015–2018	2055	2085	2055	2085
Winter Climatic Zone						
α	0.00%	0.02%	0.15%	0.25%	0.41%	0.70%
A	0.74%	2.85%	12.89%	15.97%	16.62%	42.95%
B	5.47%	8.39%	29.21%	28.28%	53.39%	51.54%
C	20.05%	37.18%	43.54%	43.03%	28.28%	4.63%
D	44.90%	49.13%	11.60%	9.84%	1.27%	0.16%
E	28.76%	2.41%	2.59%	2.60%	0.01%	0.00%
Summer Climatic Zone						
1	40.05%	8.65%	1.41%	2.11%	0.99%	0.38%
2	22.80%	23.79%	10.53%	11.83%	8.11%	0.44%
3	28.58%	38.21%	38.54%	36.45%	41.02%	13.83%
4	8.55%	29.35%	49.51%	49.60%	49.84%	85.31%
Climatic Zone (Winter + Summer)						
$\alpha3$	0.00%	0.01%	0.00%	0.00%	0.00%	0.00%
$\alpha4$	0.00%	0.01%	0.15%	0.25%	0.41%	0.70%
A1	0.00%	0.00%	0.19%	0.23%	0.11%	0.18%
A2	0.00%	0.00%	0.04%	0.08%	0.01%	0.13%
A3	0.40%	0.13%	0.03%	0.14%	0.92%	2.82%
A4	0.34%	2.72%	12.64%	15.53%	15.54%	39.74%
B1	0.00%	0.19%	0.58%	1.04%	0.33%	0.04%
B2	0.00%	0.03%	3.00%	3.09%	5.35%	0.21%
B3	3.08%	0.65%	4.04%	4.31%	22.38%	9.62%
B4	2.40%	7.52%	21.57%	19.82%	25.26%	41.63%
C1	2.87%	5.57%	0.62%	0.77%	0.41%	0.11%
C2	3.21%	3.80%	4.10%	4.86%	2.45%	0.10%
C3	8.15%	13.93%	23.80%	23.89%	16.88%	1.37%
C4	5.81%	13.85%	14.90%	13.41%	8.50%	3.05%
D1	8.21%	2.42%	0.03%	0.08%	0.13%	0.05%
D2	19.58%	17.81%	3.26%	3.68%	0.30%	0.00%
D3	16.96%	23.49%	8.06%	5.51%	0.80%	0.03%
D4	0.00%	5.26%	0.25%	0.58%	0.04%	0.09%
E1	28.76%	0.39%	0.00%	0.00%	0.01%	0.00%
E2	0.00%	2.01%	0.05%	0.04%	0.00%	0.00%
E3	0.00%	0.01%	2.54%	2.55%	0.00%	0.00%

Table 5. Dynamics of changes in energy demand between significant seasons.

	2015–2018 ED			RCP 4.5 2055 ED			Change ED between 2015–2018 and 2055		2085 ED			Change ED between 2015–2018 and 2085	
	H	C	H + C	H	C	H + C	H	C	H	C	H + C	H	C
73	64.5	10.7	75.2	32.5	13.5	46.0	32.1	−2.8	33.8	13.5	47.3	30.8	−2.8
5	8.5	24.7	33.2	5.1	31.7	36.8	3.4	−7.0	4.7	35.5	40.2	3.8	−10.8
17	28.6	17.5	46.1	11.1	23.1	34.2	17.5	−5.6	11.5	22.8	34.4	17.1	−5.4
76	14.5	22.3	36.8	10.7	29.0	39.7	3.8	−6.8	9.4	27.9	37.3	5.1	−5.6

	RCP 8.5 2055 ED			Change ED between 2015–2018 and 2055		2085 ED			Change ED between 2015−2018 and 2085	
	H	C	H + C	H	C	H	C	H + C	H	C
73	33.8	13.8	47.6	30.8	−3.1	23.5	21.4	44.9	41.0	−10.7
5	12.8	8.9	21.7	−4.3	15.8	6.4	14.1	20.5	2.1	10.6
17	−2.1	32.3	30.1	30.8	−14.8	−7.3	41.6	34.3	35.9	−24.1
76	−0.9	39.0	38.2	15.4	−16.8	−6.4	48.5	42.1	20.9	−26.2

3.1. Determination of Climate Severity Indices

Figure 2 shows the values of the new climatic severities at the location of the 77 meteorological stations, according to the CTE calculation methodology, and taking into account the climatic conditions of the period 2015–2018.

It is observed that the WCS index values range between −0.06 and 1.87, determined for the coastal province of Malaga (station #43) and Soria (station #65), respectively. The three negative WCS values recorded, one case with a value of 0, or positive but shallow values, below 0.4, have occurred in regions with mild winters; this is the case of the station above #43, located in Málaga, #15 and #42, located in Cádiz and Málaga respectively, or stations #4 (Alicante), #11 (Barcelona), #16 (Cádiz), #18 (Castellón), #28 (Huelva), #44 (Málaga), #66 (Tarragona), and #70 (Valencia), among others, all of the coastal areas in the south of the peninsula or the Mediterranean area. In these cases, a good design of the building's constructive solutions will allow the indoor comfort temperature to be reached without implementing active heating systems. On the contrary, the higher the WSC value, the lower the winter temperatures in these regions, which leads to higher heating energy demands. This is the case of stations #6 and #7 located in Avila, #13 in Burgos, #52 in Palencia, #56 in Zamora, and #64 and #65 in Soria, all of them with WSC values higher than 1.35 and located in the north of the peninsula, in provinces with shallow temperatures.

In the case of the SCS, values reaching the minimum in the coastal province of La Coruña (station #32), with 0.04, and the maximum in the inland province of Cordoba (station #20), with 2.52, are observed. The lowest SCS values occur in regions with cool summer temperatures; this is the case of stations #32, #33, #60, and #61 located in La Coruña, #34 in León, #49 and #50 in Asturias, #57 in Guipúzcoa, #58 and #59 in Cantabria, among others, all of them in the north of the peninsula and with SCS values below 0.4. In these regions, with a good design of the building's constructive solutions, the indoor comfort temperature can be reached without implementing active cooling systems. However, the higher the SCS value, the higher the summer temperatures in these regions, which leads to higher cooling energy demands. It is the case of stations located in the peninsula's interior, with SCS values higher than 2, such as #8 and #9 in Badajoz, #14 in Cáceres, #19 Ciudad Real, or #20 and #21 in Córdoba, among others.

The stations located in the inland provinces of Madrid (#40, #41), Salamanca (#55, #56), or Segovia (#62) stand out, with WCS and SCS values higher than 1. These conditions are found in places with a Mediterranean climate far from the sea, with long and cold winters, with temperatures that can drop below 0 °C, with numerous frosts occurring at night. In contrast, summers are pretty hot and dry, with a temperature range of 18.5 °C, and temperatures often exceed 30 °C. In these regions, the energy demand is considerably higher than in coastal areas, both in summer and winter, where the need for passive strategies to reduce energy consumption is essential. Thus, building solutions with high thermal inertia could be an effective mechanism to achieve thermal comfort [57].

Finally, comparing the climate severity values of the different station locations in an area shows significant temperature contrasts between the urban and metropolitan regions, for example, the case in Barcelona, where station #11, located in the city center and close to the sea coast, resulted in climatic severities of WCS = 0.23 and SCS = 1.48, while station #10, located outside the city center, showed significantly different values (with WCS = 0.61 and SCS = 0.92). The same is true for the stations in Valladolid, where station #72, located in the city center, resulted in climatic severities of WCS = 1.27 and SCS = 1.12 while station #73, located outside the city center, showed considerably different values (with WCS = 1.51 and SCS = 0.76). These results highlight the urban heat island effect, a phenomenon of thermal origin that occurs in urban areas and consists of a different temperature, which tends to be higher, especially at night, in the center of cities due to massive building [58].

3.2. Determination of the Dynamics of Changes in Climate Severity Indices

Figures 3 and 4 show the results obtained from calculating the WCS and SCS indices for the years 2055 and 2085, under the RCP 4.5 and RCP 8.5 climate change scenarios at the 77 reference stations.

For the RCP 4.5 scenario (Figure 3), it is observed that the WCS index values for 2055 range between −0.1 and 1.3, determined for the stations located in the Mediterranean coastal cities of Alicante, Barcelona, Cádiz, Castellón, Málaga, Murcia, and Valencia and the inland city of Soria, respectively. Thus, the projected emissions for this scenario mean that by mid-century, all the reference stations will be in regions with mild winters, which will lead to a drastic drop in heating consumption. It is observed that the SCS will be increasing in the first half of the 21st century, with values ranging between 0.3 and 2.8 for the cities of Oviedo (station #49) and Cordoba (station #20). It will imply a significant increase in summer temperatures, which will drastically increase cooling consumption. However, by the climate trends projected for this scenario, a progressive stabilization of temperatures at all stations will be observed by 2085. Thus, by the end of the century, WCS values range between −0.1 and 1.3; SCS values range between 0.3 and 2.9.

For the RCP 8.5 scenario (Figure 4), a more drastic change in WCS and SCS values is observed than in the previous scenario. Thus, the WCS index values for 2055 range from −0.1 to 1.4 and decrease until 2085, with values ranging from −0.1 to 1.1. In SCS, a progressive increase is observed until 2055, to increase dramatically until 2085, with values ranging between 0.2 and 3 and 0.5 and 3.7. It reflects by the end of the century and increases in cooling energy demands and an almost total decrease in heating energy demands in most cities of the reference seasons.

3.3. Proposed Update of Climate Zones for Peninsular Spain

Based on the climate severities calculated from the 77 meteorological stations for 2015–2018 and the RCP 4.5 and 8.5 scenarios, the climate severities for the period 2015–2018 and the years 2055 and 2085 for 7967 localities in peninsular Spain have been obtained using approximation techniques. Based on Table 1, the climatic zones of the 7967 localities have been identified. The results are described in Figures 5 and 6 and Tables 3 and 4 below.

3.3.1. Climate Severities for 2015–2018 and the Periods 2055 and 2085 of the RCP 4.5 and 8.5 Scenarios for 7967 Locations in Mainland Spain

Figure 5 shows the average WCS and SCS values for 2015–2018 and 2055 and 2085 for RCP 4.5 and 8.5. Comparing the values obtained for the WCS indices with those for the 2015–2018 range (Figure 5a) shows that, for the RCP 4.5 scenario, the average value decreases considerably from 0.96 in 2015–2018 to 0.57 and 0.56 in 2055 and 2085, respectively. It is because emissions in this scenario peak around 2040 and then stabilize. In the RCP 8.5 scenario, the decrease is more significant, with average values of 0.43 and 0.28 for 2055 and 2085, respectively, due to the more abrupt character of this scenario, where the most significant changes are located to the end of the 21st century. For both scenarios, there is a significant softening of winter temperatures [1].

For the SCS indices, the comparison with the values obtained in the 2015–2018 interval (Figure 5b) shows that, for the RCP 4.5 scenario, the average value increases from 1.15 in 2015–2018 to 1.43 in 2055; it then stabilizes until 2085 with 1.45. In the RCP 8.5 scenario for 2055, compared to 2015–2018, a slight increase is observed with 1.43 before rising sharply to 1.93 by 2085. The most pessimistic climate change projection will lead to more noticeable temperature changes, with sweltering summers.

Finally, it should be noted that it is not possible to compare these indices with current regulations since the CTE does not provide exact values, which leads to problems in the field of energy efficiency research and adaptation to climate change in buildings in Spain.

3.3.2. Proposed Update of Climate Zones for Peninsular Spain

Once the WCS and SCS of the 7967 localities for the period 2015–2018 and the RCP 4.8 and 8.5 scenarios have been obtained, based on Table 1, the new climate zones of the 7967 localities of peninsular Spain have been obtained. Figure 6 shows the geographical distribution of the CTE climate zones for 2015–2018 and the RCP 4.5 and 8.5 scenarios. To analyze the variation of the climate rating observed concerning the CTE, Table 3 shows the percentage of cities that vary this rating, while Table 4 shows the percentage of climate zones in the CTE in the period 2015–2018 and the RCP 4.5 and 8.5 scenarios. The results obtained are discussed below.

Proposed Update of Climate Zones for Peninsular Spain for the Period 2015–2018

Concerning the CTE in 2015–2018 (Table 3), more than 80% of the cities have already changed their climate zone (winter + summer). Moreover, this change has meant that the number of climate zones in the country has increased from the 12 contemplated in the CTE to 19, with seven new zones appearing that were not previously re-categorized (α3, α4, B1, B2, D4, E1, and E2) (Table 4). The appearance of zones α3 and α4 should be highlighted, highlighting the trend, in areas such as the Mediterranean, towards climates more characteristic of subtropical zones.

In the case of winter, approximately half of the cities have changed their winter climate zone to a warmer zone compared to the CTE (Table 3), although the most significant changes occur in the south and on the Mediterranean coast, while the climate zones in the north, northwest, southwest, and eastern part of Andalusia remain unchanged (Figure 6a,b). The winter climate zone D (Table 4) stands out, present in 49% of the localities, making it the predominant one. This result shows a rise in winter temperatures in almost half of the territory concerning what is contemplated in the current regulations. The average increase in temperatures is causing a decrease in the energy demand for heating but implies that the limits of parameters such as transmittance are compromised.

In summer climate zones, more drastic changes are observed, especially on the Mediterranean coast (Figure 6a,b), due to the intense summer warming of the Mediterranean inland waters in recent years [59]. Thus for 2015–2018, 72% of cities have changed their summer CZs to warmer ones than those reported in the CTE (Table 3). Specifically, 58.7% and 12.2% of localities have changed their summer CZ to 1 and 2 warmer ratings, respectively. Climate zones 3 and 4 are the most predominant present in 38.21% and 29.35% of the localities (Table 4).

Proposed Climate Zones for Peninsular Spain for the RCP 4.5 Scenario Projections

Under the RCP 4.5 scenario (Table 3) for 2055 and 2085, 98% of the cities will see their climate zone (winter + summer) change concerning the CTE. Furthermore, Figure 6c,d shows how the geographical distribution of climate zones for 2085 resembles the resulting distribution for 2055. These climate zone changes are a consequence of the closeness of absolute values of WCS and SCS to the limiting value of climate zone delimitation (Table 1) so that a minimal change of the index can lead to a change of climate zone for a locality. This result is a limiting factor in the zoning of the existing CTE, reinterpreting the need for a significant improvement in the development and methodology of the current regulations in force.

As shown in Figure 6c,d, by 2055 and 2085, half of the Mediterranean coastal localities will fall into climate zone A4. These regions will be characterized by hotter summers and warmer winters, while the northern coastal cities will have a greater variety of climate zones over the century, with mild summer temperatures and colder winters. The same occurs in the peninsula's interior, where a heterogeneous distribution is observed due to the complexity of the relief and the diversity of mesoclimatic and microclimatic zones. Thus, by 2085, 23.89% and 19.82% of the localities will have a C3 and B4 climate classification, while the coldest climate zones will disappear (Table 4).

In winter CZs (Table 3), for the periods 2055 and 2085, practically 90% of the localities will change their climate zoning compared to that retained in the CTE. Specifically, for the period 2055, 44.8% of the localities will change their zoning to a warmer one and 40.5% of the localities to two warmer zones. Similarly, by 2085, 41%, 43.4%, and 4.7% of localities will change their winter rating to 1, 2, and 3 warmer zones, respectively. Thus, the geographical distribution of winter climate zones for 2085 resembles the resulting distribution for 2055, where only 8% of cities will observe zone changes between these two years. The increase of the A rating concerning the CTE (Figure 6c,d and Table 4), from 0.74% in the CTE to 12.89% and 15.97% in 2055 and 2085, respectively, stands out. In contrast, the E rating decreases drastically from 28.76% in the CTE to only 2.6% in 2085. These results again show that the trend towards warmer and warmer areas will continue throughout the century.

In the summer CZ case (Table 3), for the periods 2055 and 2085, 88.6% and 87.6% of the locations will change their climate zonation compared to the CTE. By 2055, 47.7% of localities will change their zoning to a warmer one, and 39.4% of localities will change their zoning to two warmer zones. Similarly, by 2085, 48.3% and 38.1% of localities will change their summer rating to 1 and 2 warmer zones, respectively. As in the winter season, the differences between 2055 and 2085 are not significant. Although, throughout this scenario, there is a significant increase in rating 4 in the localities concerning the CTE (Figure 6c,d and Table 4), from 8.55% in the CTE to 49.51% and 49.60% in 2055 and 2085, respectively. In contrast, rating 1 decreases drastically from 40.05% in the CTE to only 2.11% in 2085. These results demonstrate the need to develop new summer zones within rating 4, with the consequent improvement in terms of building recommendations.

Proposed Climate Zones for Peninsular Spain for the RCP 8.5 Scenario Projections

Under the RCP 8.5 scenario (Table 3) for 2055 and 2085, a drastic increase in the change of climate zone classification (winter + summer) concerning the CTE is foreseen, showing that 98% of the cities will be affected. Furthermore, Figure 6e,f shows that the geographical distribution of climate zones for 2085 will undergo a significant dynamism, with zones A4 and B4, with 39.74% and 41.63%, dominating the peninsula. These zones are characterized by mild winters and sweltering summers, leading the peninsula to have climates more typical of tropical regions by the end of the century.

In the winter CZs (Table 3), for the periods 2055 and 2085, 92.4% and 100% of the locations will change their climate zoning compared to the CTE. Specifically for the period 2055, 24.4, 33.1, and 26.5% of the localities will change their qualification by one, two, and three warmer zones, respectively. Similarly, by 2085, 10.1, 36.7, and 35.4% of localities will change their winter rating to one, two, and three warmer zones, respectively. Throughout this scenario, there is a significant increase in the presence of the A and B rating in the localities concerning the CTE (Table 4), from 0.74 and 5.47% in the CTE to 42.95 and 51.54% in 2085, respectively, on the contrary, the C and D ratings decrease drastically, while the E climate zone disappears entirely by 2085. These results are due to the already indicated trend towards warmer and warmer zones.

In the summer CZ case (Table 3), for the periods 2055 and 2085, 82.7% and 91% of the localities will change their climate zoning compared to the CTE. By 2055, 33% of the localities will change their zoning to a warmer one, 25.5% of the localities to two warmer zones, and 18.9% to three warmer zones. Similarly, by 2085, 31%, 26.3%, and 32.2% of localities will change their summer rating to 1, 2, and 3 warmer zones, respectively. In Figure 6e,f, significant differences are observed between 2055 and 2085. Throughout this scenario, there is a significant increase in the presence of rating 4 in the localities concerning the CTE (Table 4), from 8.55% in the CTE to 49.84% and 85.31% in 2055 and 2085, respectively, in contrast, rating 1 decreases drastically from 40.05% in the CTE to only 0.38% in 2085. These results show that more than half of the buildings designed with zone E's technical requirements will not comply with the regulations in less than 25 years.

3.4. Analysis of the Dynamics of Dhanges in Energy Demand for RCP 4.5 and RCP 8.5 Scenarios

In order to a analyze the effects that the observed climate dynamism will have on the energy demand of buildings, Figure 7 shows the average results of the estimated change in energy demand for heating and cooling, calculated based on the definition of the WCS and SCS indices of the 77 stations, for the typical building used for 2015–2018 and the RCP 4.5 and RCP 8.5 scenarios. Besides, Table 5 shows the estimated energy demand for heating and cooling, for four significant stations, for the building type used for 2015–2018 and the RCP 4.5 and RCP 8.5 scenarios.

In the RCP 4.5 scenario (Figure 7), by the year 2055, the energy demand of the 77 stations is expected to decrease by an average of 11.23 kWh/m^2 year compared to 2015–2018. Specifically, heating demand will decrease by an average of 16 kWh/m^2 year, with the most significant decreases observed in stations located in mountainous areas at 900 meters above sea level; this is due to the notable effect of the continental climate, which will see its meteorological conditions soften in winter due to the effects of climate change. Such is the case of station 73, located in Valladolid, with a reduction of 32 kWh/m^2 year. However, cooling demand will increase by an average of 4.8 kWh/m^2 year, compared to 2015–2018, in cities located in the southeast, characterized by a semi-arid climate; specifically, station 5, located in the city of Almeria with an increase in cooling energy demand by 2055 of 7.045 kWh/m^2 year compared to 2015–2018. However, due to this climate scenario's stabilizing nature between 2055 and 2085, no significant differences are observed between the average demand values for heating and cooling, with an average increase of only 0.8 kWh/m^2 year.

Under the RCP 8.5 scenario (Figure 7), by the year 2055, heating and cooling energy demand for all seasons are expected to decrease by an average of 10 kWh/m^2 year compared to 2015–2018. Specifically, heating demand will decrease by an average of 15.4 kWh/m^2 year, highlighting station 17, located in the city of Castellón, where a high reduction value, estimated at 30 kWh/m^2 year, is foreseen. In contrast, cooling demand will increase by an average of 5.3 kW/m^2 year compared to 2015–2018. Station 76, located in Zaragoza, stands out in particular, with an increase in cooling energy demand by 2055 of 16.7 kWh/m^2 year compared to 2015–2018. It should be noted that, unlike the RCP 4.5 scenario, under the RCP 8.5 scenario, there are significant differences between 2055 and 2085. Thus, by 2085, the average heating energy demand will decrease by 8.2 kWh/m^2 year, while the cooling demand will increase by 8.7 kWh/m^2 year compared to 2055. This trend is localized in cities located in the southwestern, southern and Mediterranean coastal parts of the country. These results are explained by expanding the semi-arid climate that the south and southeast coast will undergo under this scenario, where summers will be hotter and drier, significantly increasing the energy demand of dwellings for cooling. This increase will also be affected by the additional thermal effect of the Mediterranean Sea's warming surface waters.

4. Conclusions

In this work, the climatic zones of all the cities of peninsular Spain have been updated. The results show that the allocation of climatic zones currently included in the CTE is not suitable for current and future climatic conditions. Given the importance of precision in the assignment of a climatic zone when correctly sizing domestic hot water, heating and cooling systems, and the appropriate selection of the construction materials used, this situation jeopardizes the achievement of truly sustainable buildings. Specifically, taking into account the climate data recorded in the 2015–2018 period, 80% of cities today have a different climate zone to that of the CTE; moreover, it is expected that by the year 2085 and under the forecasts recorded in the RCP 4.5 and RCP 8.5 scenarios, practically all cities in mainland Spain will change their climate zone to warmer ones.

This significant climate change that the region under study is already undergoing will help reduce the heating energy demand of dwellings and increase the demand for cooling. Therefore, architectural and construction standards must adapt to the urban environment's actual conditions and consider the main scenarios to lead to a building design that mitigates climate change and adapts to them. It intensifies the need to develop new climate zones and build recommendations to preserve future periods' correct thermal conditions.

Finally, it should be noted that the consequences observed in peninsular Spain can be extrapolated to other areas so that the methodology proposed in this work can be extended to any region, making a significant scientific contribution in terms of reflection on the current capacities and possibilities for improvement of the building stock.

Author Contributions: Conceptualization, C.D.-L., J.J., K.V., M.L.R., M.C. and M.Z.; Data curation, J.J., K.V. and M.L.R.; Formal analysis, C.D.-L., K.V. and M.Z.; Acquisition financing, M.C. and M.Z.; Research, C.D.-L., J.J., K.V., M.L.R., M.C. and M.Z.; Methodology, J.J., K.V., M.L.R. and M.Z.; Project administration, M.Z.; Resources, C.D.-L., M.C. and M.Z.; Software, J.J. and M.L.R.; Supervision, M.Z.; Validation, J.J., K.V., M.L.R. and M.Z.; Writing—original draft, C.D.-L., J.J., K.V., M.L.R., M.C. and M.Z.; Writing—revision and editing, C.D.-L., J.J., K.V., M.L.R., M.C. and M.Z. All authors have read and agreed to the published version of the manuscript.

Funding: This research was funded by the Junta de Andalucía (Research Groups FQM191, TEP-968 and FQM178), the University of Jaén (Research Structure EI_FQM8), Fundación Biodiversidad, del Ministerio para la Transición Ecológica (the Ministry for Ecological Transition) Agencia Nacional de Investigación y Desarrollo (ANID) de Chile: ANID FONDECYT 1201052: ANID PFCHA/DOCTORADO BECAS CHILE/2019—21191227.

Institutional Review Board Statement: Not applicable.

Informed Consent Statement: Not applicable.

Data Availability Statement: Data are provided upon request to the corresponding author.

Conflicts of Interest: The authors declare no conflict of interest.

Abbreviations

AdapteCCa	Climate Change Adaptation Platform
AEMET	State Meteorological Agency
C	cooling energy demand
CDD	cooling degree-days
CSI	climatic severity index
CTE	Technical Building Code
CZ	climate zone
DB-HE	Basic Document on Energy Saving
EC	energy consumption
ED	energy demand
ESRL	Earth System Research Laboratory
GDP	Gross Domestic Product

H	heating energy demand
HDD	heating degree-days
IPCC	Intergovernmental Panel on Climate Change
LLGHG	long-lived greenhouse gas emissions
RBF	radial basis functions
RF	Radiative Forcing
RCP	Representative Concentration Pathways
SCS	severity index for summer
WCS	severity index for winter
WS	Weather station

References

1. IPCC. *Climate Change 2014: Synthesis Report. Contribution of Working Groups I, II and III to the Fifth Assessment Report of the Intergovernmental Panel on Climate Change*; Core Writing Team, Pachauri, R.K., Meyer, L.A., Eds.; IPCC: Geneva, Switzerland, 2014; 151p.
2. Houghton, J.T.; Ding, Y.D.J.G.; Griggs, D.J.; Noguer, M.; van der Linden, P.J.; Dai, X.; Johnson, C.A. *Climate Change 2001: The Scientific Basis*; The Press Syndicate of the University of Cambridge: Cambridge, UK, 2001; p. 881. [CrossRef]
3. Chuwah, C.; Van Noije, T.; Van Vuuren, D.P.; Hazeleger, W.; Strunk, A.; Deetman, S.; Beltran, A.M.; Van Vliet, J. Implications of Alternative Assumptions Regarding Future Air Pollution Control in Scenarios Similar to the Representative Concentration Pathways. *Atmos. Environ.* **2013**, *79*, 787–801. [CrossRef]
4. Verichev, K.; Zamorano, M.; Carpio, M. Effects of Climate Change on Variations in Climatic Zones and Heating Energy Consumption of Residential Buildings in the Southern Chile. *Energy Build.* **2020**, *215*, 109874. [CrossRef]
5. Thomson, A.M.; Calvin, K.V.; Smith, S.J.; Kyle, G.P.; Volke, A.; Patel, P.; Delgado-Arias, S.; Bond-Lamberty, B.; Wise, M.A.; Clarke, L.E.; et al. RCP 4.5: A Pathway for Stabilization of Radiative Forcing by 2100. *Clim. Chang.* **2011**, *109*, 77–94. [CrossRef]
6. Riahi, K.; Rao, S.; Krey, V.; Cho, C.; Chirkov, V.; Fischer, G.; Kindermann, G.; Nakicenovic, N.; Rafaj, P. RCP 8.5—A Scenario of Comparatively High Greenhouse gas Emissions. *Clim. Chang.* **2011**, *109*, 33–57. [CrossRef]
7. Lorusso, A.; Maraziti, F. Heating System Projects Using the Degree-Days Method in Livestock Buildings. *J. Agric. Eng. Res.* **1998**, *71*, 285–290. [CrossRef]
8. Troup, L.; Eckelman, M.J.; Fannon, D. Simulating Future Energy Consumption in Office Buildings Using an Ensemble of Morphed Climate Data. *Appl. Energy* **2019**, *255*, 113821. [CrossRef]
9. Bellia, L.; Mazzei, P.; Palombo, A. Weather Data for Building Energy Cost-Benefit Analysis. *Int. J. Energy Res.* **1998**, *22*, 1205–1215. [CrossRef]
10. Grøntoft, T. Climate Change Impact on Building Surfaces and Façades. *Int. J. Clim. Chang. Strat. Manag.* **2011**, *3*, 374–385. [CrossRef]
11. Nik, V.M.; Mundt-Petersen, S.; Kalagasidis, A.S.; De Wilde, P. Future Moisture Loads for Building Facades in Sweden: Climate Change and Wind-Driven Rain. *Build. Environ.* **2015**, *93*, 362–375. [CrossRef]
12. Brown, M.A.; Cox, M.; Staver, B.; Baer, P. Modeling climate-driven changes in U.S. buildings energy demand. *Clim. Chang.* **2016**, *134*, 29–44. [CrossRef]
13. Christenson, M.; Manz, H.; Gyalistras, D. Climate warming impact on degree-days and building energy demand in Switzerland. *Energy Convers. Manag.* **2006**, *47*, 671–686. [CrossRef]
14. De Rosa, M.; Bianco, V.; Scarpa, F.; Tagliafico, L.A. Heating and cooling building energy demand evaluation; a simplified model and a modified degree days approach. *Appl. Energy* **2014**, *128*, 217–229. [CrossRef]
15. Dolinar, M.; Vidrih, B.; Kajfež-Bogataj, L.; Medved, S. Predicted changes in energy demands for heating and cooling due to climate change. *Phys. Chem. Earth Parts A/B/C* **2010**, *35*, 100–106. [CrossRef]
16. Jylhä, K.; Jokisalo, J.; Ruosteenoja, K.; Pilli-Sihvola, K.; Kalamees, T.; Seitola, T.; Mäkelä, H.M.; Hyvönen, R.; Laapas, M.; Drebs, A. Energy demand for the heating and cooling of residential houses in Finland in a changing climate. *Energy Build.* **2015**, *99*, 104–116. [CrossRef]
17. Berardi, U.; Jafarpur, P. Assessing the impact of climate change on building heating and cooling energy demand in Canada. *Renew. Sustain. Energy Rev.* **2020**, *121*, 109681. [CrossRef]
18. da Guarda, E.L.A.; Domingos, R.M.A.; Jorge, S.H.M.; Durante, L.C.; Sanches, J.C.M.; Leão, M.; Callejas, I.J.A. The influence of climate change on renewable energy systems designed to achieve zero energy buildings in the present: A case study in the Brazilian Savannah. *Sustain. Cities Soc.* **2020**, *52*, 101843. [CrossRef]
19. Hekkenberg, M.; Moll, H.; Uiterkamp, A.S. Dynamic temperature dependence patterns in future energy demand models in the context of climate change. *Energy* **2009**, *34*, 1797–1806. [CrossRef]
20. Seljom, P.; Rosenberg, E.; Fidje, A.; Haugen, J.E.; Meir, M.; Rekstad, J.; Jarlset, T. Modelling the effects of climate change on the energy system—A case study of Norway. *Energy Policy* **2011**, *39*, 7310–7321. [CrossRef]
21. Belzer, D.B.; Scott, M.J.; Sands, R.D. Climate Change Impacts on U.S. Commercial Building Energy Consumption: An Analysis Using Sample Survey Data. *Energy Sources* **1996**, *18*, 177–201. [CrossRef]

22. Amato, A.D.; Ruth, M.; Kirshen, P.; Horwitz, J. Regional Energy Demand Responses to Climate Change: Methodology and Application to the Commonwealth of Massachusetts. *Clim. Chang.* **2005**, *71*, 175–201. [CrossRef]
23. Xu, P.; Huang, Y.J.; Miller, N.; Schlegel, N.; Shen, P. Impacts of climate change on building heating and cooling energy patterns in California. *Energy* **2012**, *44*, 792–804. [CrossRef]
24. Asimakopoulos, D.; Santamouris, M.; Farrou, I.; Laskari, M.; Saliari, M.; Zanis, G.; Giannakidis, G.; Tigas, K.; Kapsomenakis, J.; Douvis, C.; et al. Modelling the Energy Demand Projection of the Building Sector in Greece in the 21st Century. *Energy Build.* **2012**, *49*, 488–498. [CrossRef]
25. Olonscheck, M.; Holsten, A.; Kropp, J.P. Heating and Cooling Energy Demand and Related Emissions of the German Residential Building Stock under Climate Change. *Energy Policy* **2011**, *39*, 4795–4806. [CrossRef]
26. Zhai, Z.J.; Helman, J.M. Implications of Climate Changes to Building Energy and Design. *Sustain. Cities Soc.* **2019**, *44*, 511–519. [CrossRef]
27. Shi, Y.; Wang, G. Changes in Building Climate Zones over China Based on High-Resolution Regional Climate Projections. *Environ. Res. Lett.* **2020**, *15*, 114045. [CrossRef]
28. Belcher, S.; Hacker, J.; Powell, D. Constructing Design Weather Data for Future Climates. *Build. Serv. Eng. Res. Technol.* **2005**, *26*, 49–61. [CrossRef]
29. Dodoo, A.; Gustavsson, L.; Bonakdar, F. Effects of Future Climate Change Scenarios on Overheating Risk and Primary Energy Use for Swedish Residential Buildings. *Energy Procedia* **2014**, *61*, 1179–1182. [CrossRef]
30. Roux, C.; Schalbart, P.; Assoumou, E.; Peuportier, B. Integrating Climate Change and Energy Mix Scenarios in Lca Of Buildings and Districts. *Appl. Energy* **2016**, *184*, 619–629. [CrossRef]
31. Sailor, D.J. Risks of Summertime Extreme Thermal Conditions in Buildings as a Result of Climate Change and Exacerbation of Urban Heat Islands. *Build. Environ.* **2014**, *78*, 81–88. [CrossRef]
32. Guan, L. Preparation of Future Weather Data to Study the Impact of Climate Change on Buildings. *Build. Environ.* **2009**, *44*, 793–800. [CrossRef]
33. Wang, X.; Chen, D.; Ren, Z. Assessment of Climate Change Impact on Residential Building Heating and Cooling Energy Requirement in Australia. *Build. Environ.* **2010**, *45*, 1663–1682. [CrossRef]
34. Andrić, I.; Gomes, N.; Pina, A.; Ferrão, P.; Fournier, J.; Lacarrière, B.; Le Corre, O. Modeling the Long-Term Effect Of Climate Change on Building Heat Demand: Case Study on a District Level. *Energy Build.* **2016**, *126*, 77–93. [CrossRef]
35. EUR-Lex-32018L0844-EN-EUR-Lex. Available online: https://eur-lex.europa.eu/eli/dir/2018/844/oj (accessed on 15 April 2020).
36. Mulvaney, D. Green New Deal. *Solar Power* **2019**, 47–65. [CrossRef]
37. Australian Building Codes Board. Handbook: Energy Efficiency NCC Volume Two. 2019. Available online: https://www.abcb.gov.au/Resources/Publications/Education-Training/energy-efficiency-ncc-volume-two (accessed on 14 January 2021).
38. International Code Council. *The International Energy Conservation Code (IECC)*. 2000. Available online: https://shop.iccsafe.org/2000-international-energy-conservation-coder-pdf-download.html (accessed on 22 December 2020).
39. HULC Unified Tool LIDER-CALENER, Version 1.0.1493.1049 (Herramienta Unificada LIDER-CALENER, Versión 1.0.1493.1049). Available online: http://www.codigotecnico.org/index.php/menu-recursos/menu-aplicaciones/282-herramienta-unificada-lider-calener (accessed on 1 July 2016).
40. Rakoto-Joseph, O.; Garde, F.; David, M.; Adelard, L.; Randriamanantany, Z. Development of Climatic Zones and Passive Solar Design in Madagascar. *Energy Convers. Manag.* **2009**, *50*, 1004–1010. [CrossRef]
41. Walsh, A.; Cóstola, D.; Labaki, L.C. Review of Methods for Climatic Zoning for Building Energy Efficiency Programs. *Build. Environ.* **2017**, *112*, 337–350. [CrossRef]
42. Ministerio de Fomento Documento Básico HE Ahorro de Energía 2019. *Código Técnico de la Edificacion* **2019**, 1–129.
43. Ferreira, M.; Coelho, M.J.P.; Alves, R.V.L. *Regulamento das Características de Comportamento Térmico de Edifícios (RCCTE): Desenvolvimento de Folha de Cálculo*; Edições Universidade Fernando Pessoa: Porto, Portugal, 2009; pp. 67–75.
44. France. Code de La Construction et de l' Habitation Partie Législative Livre Ier: Dispositions Générales. Titre Préliminaire: Informations Du Parlement En Matière de Logement Partie Législative Livre Ier: Dispositions Générales. *Titre Ier Constr. Des.* **2011**, 2007–2008.
45. Carpio, M.; Jódar, J.; Rodíguez, M.L.; Zamorano, M. A Proposed Method Based on Approximation and Interpolation for Determining Climatic Zones and Its Effect on Energy Demand and CO_2 Emissions from Buildings. *Energy Build.* **2015**, *87*, 253–264. [CrossRef]
46. Lastra-Bravo, X.B.; Fernández-Membrive, V.J.; Flores-Parra, I.; Tolón-Becerra, A. Energy Qualification in Buildings. The Effect of Changes in Construction Methods in the Spanish A4 Climate Zone. *Energy Procedia* **2013**, *42*, 513–522. [CrossRef]
47. Salmerón, J.; Álvarez, S.; Molina, J.; Ruiz, A.; Sánchez, F. Tightening the Energy Consumptions of Buildings Depending on Their Typology and on Climate Severity Indexes. *Energy Build.* **2013**, *58*, 372–377. [CrossRef]
48. State Meteorological Agency-AEMET-Spanish Government. Available online: http://www.aemet.es/es/portada (accessed on 15 April 2021).
49. NOAA ESRL GMD NOAA Solar Calculation. Available online: https://www.esrl.noaa.gov/gmd/grad/solcalc/calcdetails.html (accessed on 20 January 2021).

50. National Platform for Adaptation to Climate Change. Available online: https://www.adaptecca.es/en (accessed on 20 January 2021).
51. Jiang, A.; Zhu, Y.; Elsafty, A.; Tumeo, M. Effects of Global Climate Change on Building Energy Consumption and Its Implications in Florida. *Int. J. Constr. Educ. Res.* **2018**, *14*, 22–45. [CrossRef]
52. Chan, A. Developing Future Hourly Weather Files for Studying the Impact of Climate Change on Building Energy Performance in Hong Kong. *Energy Build.* **2011**, *43*, 2860–2868. [CrossRef]
53. Burden, R.L.; Faires, J.D.; Burden, A.M. *Numerical Analysis*; Cengage: New York, NY, USA, 2016; ISBN 9781305253667.
54. Buhmann, M.D. Radial basis functions. *Acta Numer.* **2000**, *9*, 1–38. [CrossRef]
55. López-Ochoa, L.M.; Las-Heras-Casas, J.; López-González, L.M.; García-Lozano, C. Environmental and Energy Impact of the Epbd in Residential Buildings in Cold Mediterranean Zones: The Case of Spain. *Energy Build.* **2017**, *150*, 567–582. [CrossRef]
56. Royal Decree 314/2006 of March 17, Approving the Technical Building Code (Real Decreto 314/2006, de 17 de Marzo, Por El Que Se Aprueba El Código Técnico de La Edificación). Available online: http://www.boe.es/boe/dias/2006/03/28/pdfs/A11816-11831.pdf (accessed on 20 July 2020).
57. Avendaño-Vera, C.; Martinez-Soto, A.; Marincioni, V. Determination of Optimal Thermal Inertia of Building Materials for Housing in Different Chilean Climate Zones. *Renew. Sustain. Energy Rev.* **2020**, *131*, 110031. [CrossRef]
58. Parker, J. The Leeds Urban Heat Island and Its Implications for Energy Use and Thermal Comfort. *Energy Build.* **2021**, *235*, 110636. [CrossRef]
59. Adloff, F.; Somot, S.; Sevault, F.; Jordà, G.; Aznar, R.; Déqué, M.; Herrmann, M.; Marcos, M.; Dubois, C.; Padorno, E.; et al. Mediterranean Sea Response to Climate Change in an Ensemble of Twenty First Century Scenarios. *Clim. Dyn.* **2015**, *45*, 2775–2802. [CrossRef]

applied sciences

MDPI

Review

Cool Surface Strategies with an Emphasis on the Materials Dimension: A Review

Chaimae Mourou [1], Montserrat Zamorano [2], Diego P. Ruiz [3] and María Martín-Morales [1,*]

[1] Department of Building Construction, School of Building Engineering, University of Granada, 18071 Granada, Spain; chaimaemourou@correo.urg.es
[2] Department of Civil Engineering, School of Civil Engineering, University of Granada, 18071 Granada, Spain; zamorano@ugr.es
[3] Department of Applied Physics, Faculty of Sciences, University of Granada, 18071 Granada, Spain; druiz@ugr.es
* Correspondence: mariam@ugr.es

Abstract: The need to tackle the urban heat island effect demands the implementation of cool surfaces as a mitigation strategy. This study comprehensively reviews the evolution of this research field from a materials perspective. It provides a bibliometric analysis of the relevant literature using the SciMAT software processing of bibliographic records from 1995 to 2020, for the evolution of cool surfaces. The results obtained show an increased interest in the field from 2011 to 2020, particularly for roof applications, and present the scientific evolution of reflective materials. According to the materials dimension adopted by the development of the research field, the study is refined from a bibliometric analysis of 982 selected records for the analysis of five themes: (i) Pigments; (ii) Phase change materials; (iii) Retroreflective materials; (iv) Ceramic materials; and (v) Glass. These materials present promising results in terms of their solar reflectance performances in the mitigation of the urban heat island phenomenon. At the end of this review, recommendations for future studies are provided for the creation of economic and environmentally friendly materials based on waste glass recycling. This study represents a valuable contribution that provides a scientific background with regard to cool surfaces from a materials perspective for future investigations.

Keywords: cool surface; cool material; cool roof; urban heat island

Citation: Mourou, C.; Zamorano, M.; Ruiz, D.P.; Martín-Morales, M. Cool Surface Strategies with an Emphasis on the Materials Dimension: A Review. *Appl. Sci.* **2022**, *12*, 1893. https://doi.org/10.3390/app12041893

Academic Editor: Elza Bontempi

Received: 7 January 2022
Accepted: 5 February 2022
Published: 11 February 2022

Publisher's Note: MDPI stays neutral with regard to jurisdictional claims in published maps and institutional affiliations.

1. Introduction

Urban areas represent 2% of the Earth's surface, yet they consume 75% of the world's energy resources [1]. A portion of this energy is dissipated as anthropogenic heat, which increases the ambient temperatures in urban areas. This phenomenon was labelled as the "urban heat island (UHI) effect" by meteorologists more than a century ago, and it is the result of the increase in the ambient temperatures and the amount of solar heat trapped in urban areas, as well as of the increase in greenhouse gas emissions [2–4]. Consequently, in addition to climate change, associated environmental challenges and public health problems have occurred [4]; for instance, the occurrence of urban smog and the increase in cooling energy consumption [2], as well as a decrease in indoor and outdoor thermal comfort and quality [3,5]. It can therefore be stated that the UHI effect is an environmental problem that requires theoretical and practical studies to mitigate its impact [6].

In order to palliate the UHI effect, a growing number of studies and investigations have been conducted to develop mitigation strategies that can be implemented in urban spaces and buildings. Several solutions have been developed, including urban geometry re-shaping [1], designing green and cool roofs [7,8], using permeable, porous, water-retentive, and cool pavements [9], incorporating green spaces into the urban landscape, and utilizing water and wind for cooling effects [4,10,11]. These solutions could yield a median reduction in the air temperature of between 1.8 and 2.1 K [12]. Furthermore, the combination of different measures could be more effective [13], and the choice of the optimal strategy depends on the regional atmospheric and geographic specifications of each individual urban environment [8,14].

Among the cited UHI mitigation strategies, cool surfaces have emerged as a viable solution [15,16]. The term "cool surfaces" basically refers to surfaces with reflective materials and coatings that reflect the solar energy radiation that hits building envelopes and urban areas [17], including roofs, facades, and pavements (Figure 1). Cool surfaces are able to reduce the thermal infrared radiation outflow in the atmosphere, as well as the temperature and the solar heat gain [18,19]. In fact, it has been proven that the implementation of cool surfaces to replace dark and highly absorptive materials during routine maintenance increases the albedo over time [2]. These materials come in a huge variety and include natural materials, artificial cool coatings, and nonwhite high-albedo materials [20,21]. In addition to the fabrication process and conditions, the thickness, particle size, and the substrate and binder materials are all key parameters that could affect the optical and thermal properties of cool materials, such as the albedo, permeability, conductivity, total solar reflectance, and emissivity [21,22].

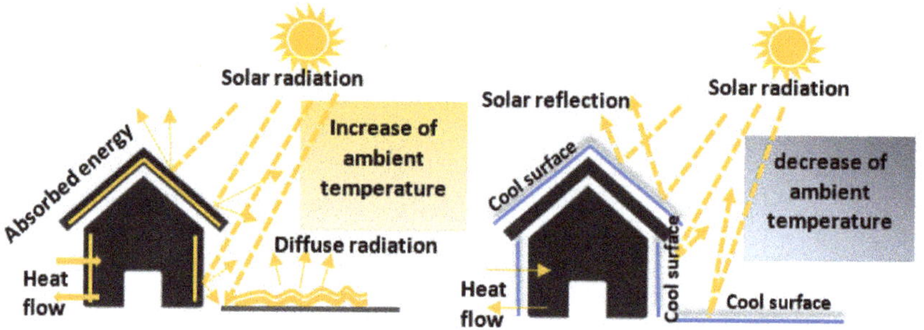

Figure 1. Schematic representation of the cool surface effect on a UHI through the ambient temperature.

The thermal performance of a material is mainly evaluated by the albedo (solar radiation) and emissivity (longwave radiation) [21]. The solar reflectance potential of cool materials initially relied on their whiteness, which promoted the use of white paints and light-colored aggregates [23]. Following that, the research field advanced toward enhancing the near infrared (NIR) reflectance of cool materials, as it represents almost 52% of the electromagnetic spectrum of light (i.e., from 700 to 2500 nm) [24]. Thus, novel methodologies and techniques have been considered to cover a wide range of solar reflectances in order to enhance the performance of cool materials.

The cool surface strategy has started an extensive series of studies concerning pigments that has created an industry of geoengineering and chemical solutions for the development of cool materials to enhance the solar reflectance performances. The integration of pigments in cool materials has recently been discussed in detail within both the research and industrial contexts [25]. With regard to the aesthetic requirements of a design, selective pigments have been developed to maintain the optical color desired on top of the material while achieving important NIR reflectance results [26]. The pigments range from organic, to complex inorganic color mixed-metal oxides [27,28]. These substances have demonstrated high solar reflectances of up to 95%, compared to TiO_2 [29–33].

In addition to cool pigments, other solutions for cool surfaces have emerged to improve urban climate conditions and energy consumption, such as retroreflective (RR) and phase change materials (PCMs). Independently of the incidence direction, retroreflectivity refers to the capacity of a surface to reflect an incoming light beam to a surface back towards its source [34,35]. RR materials have been demonstrated to be effective in several studies in terms of the solar radiation reflectance beyond urban canyons and canopies [34–36]. PCMs have the ability to change their physical characteristics as a consequence of heat release or absorption [37]. In recent decades, a variety of PCMs have been investigated as dynamic components in structures [38]. The implementation of PCMs in the matrix of roof finishing materials decreases the flux of the roof heat gain by 54%, compared to the cool roof [16], and it helps to compensate for the effect of the thermal stress generated by the latter [39]. Moreover, PCMs can be used to regulate the indoor thermal comfort and reduce the heating penalty during the summer and winter, respectively, more so than do cool paints [40–43].

Sustainable adaptation has been a parallel concern in the development of cool material solutions. On this basis, the use of recycling materials to save energy and natural resources, and to enhance the solar reflectance, presents a promising eco-friendly strategy [14]. For instance, full body porcelain aggregate from waste tiles has been used as a cool pavement coating that exhibits important thermal performance values compared to asphalt pavement, with an NIR solar reflectance of 6.4 °C, and a surface temperature reduction of up to 6.4 °C during the peak periods [44]. In addition, the use of recycled glass cullet in the fabrication of a sustainable asphalt roof shingle improved the solar reflectance [45].

Considering the evolution of materials applied in cool surfaces, the current objective of this study is to perform a bibliometric analysis to review the scientific development of this solution in the mitigation of the UHI effect. Section 2 presents the research methodology employed. Section 3 describes the collection process for the materials and the bibliometric evaluation, provides a descriptive analysis of the findings and analyzes the most important contributions of the studies identified previously. Finally, Section 4 provides the most important conclusions of this study, which contribute to the existing body of knowledge by highlighting the trends in the research field of cool surfaces for building envelopes and urban areas, as well as by recommending research areas for future studies.

2. Methodology

To achieve the objective of this study, a dual bibliometric study based on science mapping and performance analysis was conducted first using the science mapping analysis software, SciMAT, to obtain the necessary patterns and bibliometric measures [46,47]. This bibliometric study contains publications from 1995 to 2020, and it was conducted in 2021. Hence, recent findings and limitations are included in the literature review section in order to enrich the discussion, which is necessary because of the constant evolution of the field. Science mapping visualizes, analyzes, and models a broad range of scientific and technological activities, and it follows a general workflow of data retrieval, data pre-processing, network extraction, network normalization, mapping, analysis, visualization, and finally, the interpretation of the results [46]. Furthermore, the performance analysis as a complementary methodology uses different bibliometric measures and indicators to complement the visualization results and to help identify the impacts and productivities of the themes in the research field. With consideration to the results obtained, a number of publications were selected in order to conduct a review of the evolution of cool surfaces.

2.1. Sample Definition and Steps for the Data Collection

This review addresses two concepts: the UHI effect, and the different types of cool surfaces applied. To achieve this, an extensive search was carried out that employed the keywords linked to both concepts, as well as the satellite materials directly connected to the research field. Therefore, the first stage of the data collection was performed using the field, "Title/Abstract/Keyword", through the following keywords and search strings (Figure 2): "Heat island" OR "Reflect*" AND "Cool surface*", "Cool facade*", and "Cool roof*" AND "Cool pavement*", within the Web of Science Core Collection and Scopus databases. The asterisk is used at the end of keywords to broaden the research. As a result, 982 publications were found, 347 of which were excluded after the deduplicating and cleaning of the raw data. After reading the abstracts, another 121 publications were excluded because they were not aligned with the purpose of this research. Finally, the bibliometric study was performed with 514 publications.

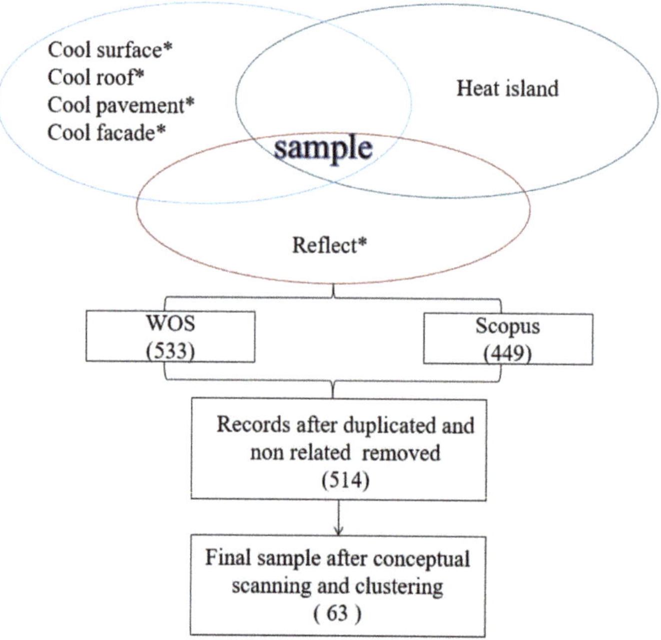

Figure 2. Data collection flowchart.

The next step was dedicated to the restriction of the data and the refinement of the sample on the basis of a conceptual approach. According to the SciMAT visualization results, it was possible to follow the internal links in the cluster networks of the concepts with the highest potentials for reviewing, which returned 63 records. This process of data collection is described in the flowchart above (Figure 2). The analysis of these publications was the basis of the literature review of the evolution of cool surfaces through the materials applied to mitigate the UHI effect.

2.2. Systematic Literature Review

The analysis of the 514 documents selected for the systematic literature review resulted in the following data: total numbers and years of publications; and authors with the highest contributions to the field based on the number of publications, sources, and journals. Furthermore, strategic diagrams, thematic networks, overlay graphs, and evolution maps were used to demonstrate the relationships between each theme of the strategic diagrams, the keywords, and their interconnections. These were also used to identify the research motor themes, the highly developed and isolated themes, the emerging or declining themes, and the basic and transversal themes, and to then trace the evolution of these themes along the studied period [48–50]. The sample of academic publications processed through this science mapping analysis was dependent on the specific input conditions, such as the unit of analysis and the keywords.

3. Results and Discussion

The aim of this section is to assess and analyze the collected material that was released during the period between 1995 and 2020 using quantitative and qualitative methods. A descriptive analysis was conducted through: (i) An analysis of the evolution of the numbers of documents; and (ii) The main sources of the publications and a review of the most prolific authors. Science mapping and visualization were then conducted to obtain an assessment of the evolution of the cool-material-application domain. Finally, a literature revision of the materials applied for cool surfaces was developed, which led to five analytical sections employed for the evaluation of the materials: (i) Pigments; (ii) RR materials; (ii) PCMs; (iv) Ceramic materials; and (v) Glass.

3.1. Descriptive Analysis
3.1.1. Evolution of Number of Documents

Since the first article identified in these databases was published in 1995, the time horizon used in this study was from 1995 to 2020. In order to analyze the trends and patterns in the publications, three periods were identified (1995–2001, 2002–2010, and 2011–2020) on the basis of the main turning points and milestones in the evolution of cool surfaces. The results of the contributions for the three periods identified are summarized below.

The first period (1995–2001): In this period, the contribution of the academic literature to the topic was very poor, with only five publications found. However, the period was marked by promoting cool roofs through building codes. In fact, starting in 1999, several energy-building standards adopted cool roof credits or requirements, such as ASHRAE 90.1, ASHRAE 90.2, the International Energy Conservation Code, and California's Title 24.

The second period (2002–2010): This period is characterized by the rising concern with regard to the topic, which coincided with the third assessment report of the Intergovernmental Panel on Climate Change, which highlights the increase in greenhouse emissions and the global average surface temperature in the 20th century by 0.6 °C. The number of publications increased considerably (to 58) during this period (Figure 3).

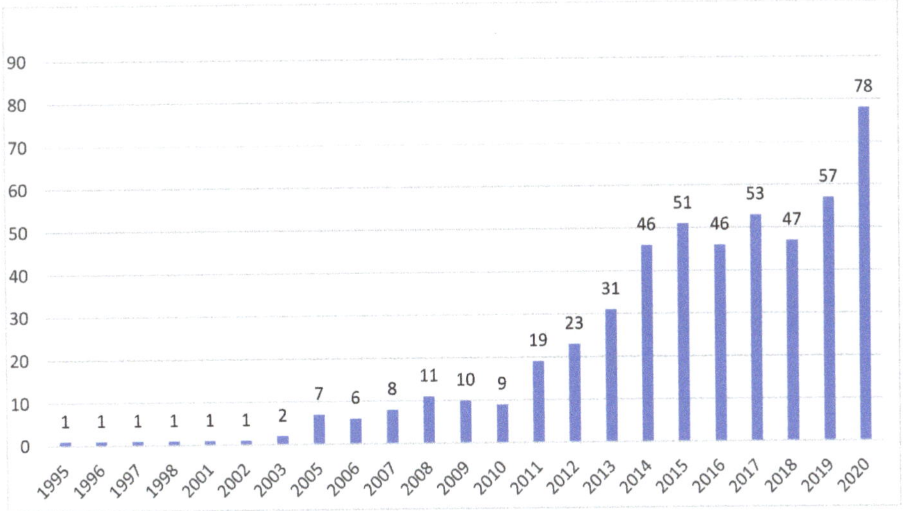

Figure 3. Number of documents per year.

The third period (2011–2020): The cool surface strategy was a rapidly growing research topic during this period. In fact, this period is by far the most prolific in terms of publications, with 87.74% of the records published during this time (Figure 3). This period coincides with the foundation of the European Cool Roofs Council, which seeks to develop knowledge and research regarding cool roof technology and promotes the use and implementation of this technology in Europe.

Considering the significant evolution of the number of academic documents in this field, an increasing interest in this topic has been explicitly detected (Figure 3).

3.1.2. Main Source Publications

The nine most prolific journals account for 40% of the total records of the sample principally considered. It is possible to conclude that the most common aspect of the journals considered is related to energy efficiency and buildings, as well as to the science and technology of solar energy applications. In fact, *Energy and Buildings* has the dominant share, with 16.53% of the published records, followed by *Solar Energy* (6.19%), and *Solar Energy Materials and Solar Cells* (4.12%).

Concerning the authorship of the publications, Akbari, H., who is affiliated with the United States and Canada, has made the most contributions to the field, with 36 publications, followed by Pisello, A.L. (33), and Levinson, R. (28). Most of the publications are affiliated with Italian universities (97 papers), followed by the United States and Canada, with 74 and 36 publications, respectively.

During the last decade, with the increases in the UHI effect and climate change, it became necessary to draw attention to the development of geoengineering-based solutions, such as cool surfaces. This strategy has been implemented for roof applications, building facades, and pavements. Nevertheless, the academic research placed an emphasis on cool roofs, with a much higher number of articles published on this application than for any of the others (Figure 4). Thus, the number of publications regarding cool roofs in the third period reached 251 documents, while, in the case of pavements and facades, only 58 and 22 papers were identified, respectively.

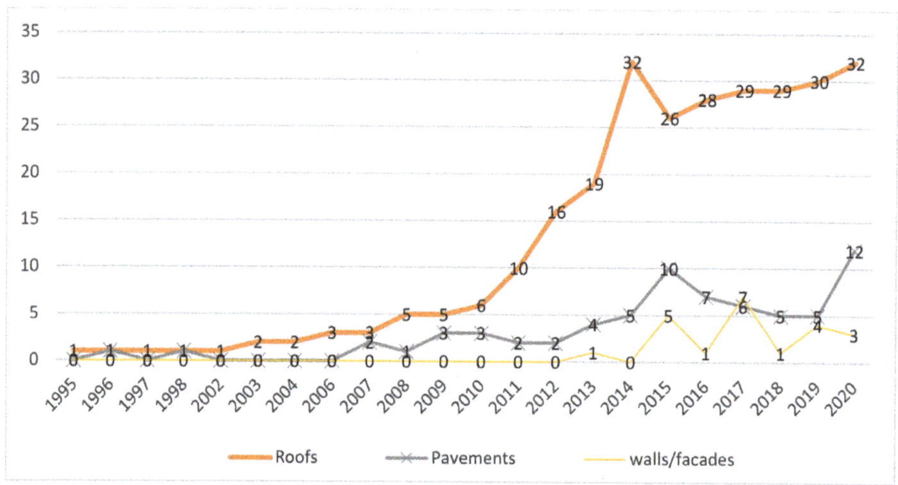

Figure 4. Evolution of cool material application domain: roofs, pavements, and walls/facades.

3.2. Science Mapping and Visualization

Figure 5 plots the overlapping map representing the three periods of research and the evolution of the keywords. Figure 6 presents the thematic evolution map of the research field, based on the h-index (Figure 6a) and on the numbers of published documents for the three cited periods (Figure 6b). Figure 5 shows that the number of keywords grew substantially in the second period, with 52 units, and in the third period, with 16 more. Thus, it is possible to confirm that cool surfaces represent a growing research field. In fact, the first period was the least developed in terms of published documents, with most of them focused on the use of lighter-colored materials to provide high solar energy reflectance and to decrease the solar heat entrapped in urban areas [2,23]. This period was characterized by the "cooling" theme (Figure 6b), as the increases in the temperatures in cities urged the use of high-albedo surfaces instead of darker materials.

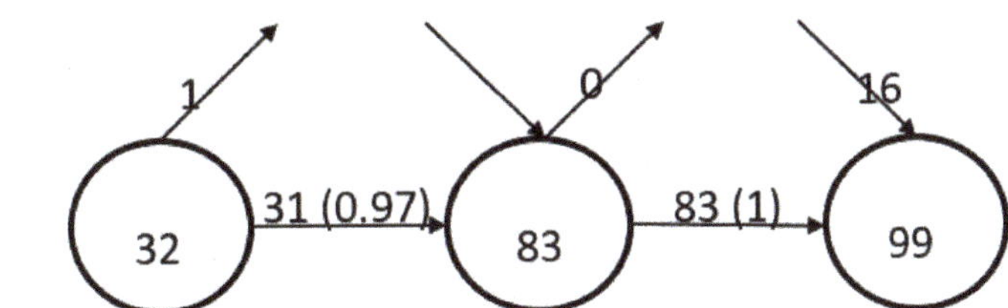

Figure 5. Overlapping map of the sample.

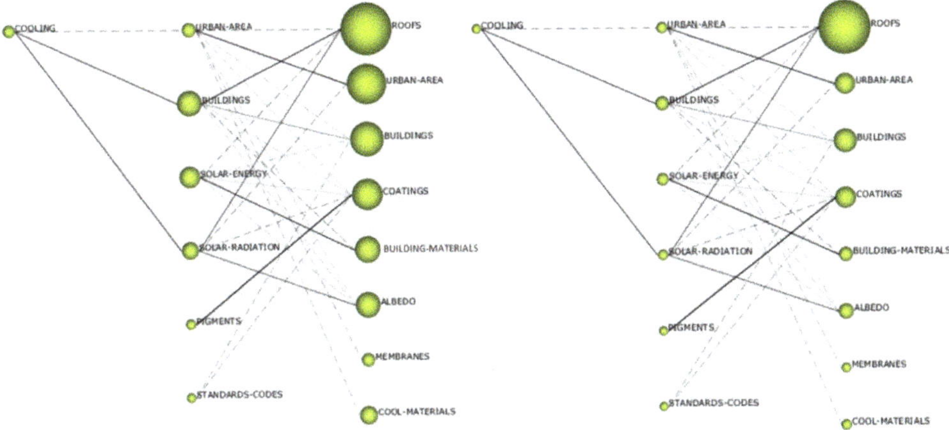

a. Thematic evolution map based on the h-index

b. Thematic evolution map based on the number of published documents

Figure 6. (Thematic evolution map according to the h-index (**a**) and the number of published documents (**b**).

Later, several studies showed that the whiteness of materials enhances the surface albedo and increases the solar reflectance in the visible spectrum range, which reduces the cooling load of buildings [51]. For instance, in a study exploring alternative methods for creating high-albedo concrete for pavement applications, Boriboonsomsin and Reza [52] found that replacing cement with whiter constituents (70% slag) achieves an albedo of 0.582, which is 71% higher than the conventional mix. To encourage the implementation of white reflective materials in buildings and urban areas, as well as to facilitate their integration into the construction sector, standards and product labelling were adopted and promoted on the basis of the spectral reflection examination of these materials [2,53,54]. In this sense, some efforts have been taken to incorporate cool roofs as an effective sustainable strategy in the revised ASHRAE building standards, S90.1 [55]. This approach was developing in the second period and it coincides with the integration of building regulations to enhance energy performances, such as the first version of the Energy Performance of Buildings Directive, 2002/91/EC, as well as its subsequent update (Directive 2010/31/EU). In this second period, the research field started to receive more interest, which is highlighted by the inclusion of 52 new keywords (Figure 5) and the following six emerging themes (Figure 6): "buildings"; "solar-energy"; "solar-radiation"; "urban-area"; "pigments"; and "standards-codes". The h-index impact in Figure 6a emphasizes buildings, solar energy, and solar radiation, and the number of published documents is approximately the same for each theme (Figure 6b). These studies began because of the rising interest in cool materials with solar radiation reflectance properties not only in the visible range, but also in the NIR spectrum, to reduce energy consumption and enhance thermal comfort.

Finally, the third period increases the number of keywords to 99 (Figure 5), and it shows a link between the research field and the creation of balanced solutions in the development of coatings, membranes, and materials designed to save energy in buildings (Figure 6b). This represents the most prolific period in terms of published documents. In this period, the conceptual evolution of the themes was developed into more specific areas directed toward creating solutions and developing materials, such as: "roofs"; "urban-area"; "buildings"; "coatings"; "building-materials"; "albedo"; "membranes"; and "cool-materials". Researchers started to explore more alternatives that conformed to the energy efficiency and aesthetic requirements, both indoors and outdoors, and there was increasing interest in materials with the appropriate thermal emissivity/absorption spectrum. For this purpose, the careful selection of nanoparticles and pigments was developed to optimize the thermal and optical performances of materials, such as radiative cooling painting. Several works were led in this sense to develop high-reflective paints known as "cool paints", such as smart coatings with high NIR reflectance to reduce the solar heat gain and the energy consumption [56–59]. In addition, the coatings were developed in order to achieve the color and efficient radiative cooling requirements in a simple, low-cost, and scalable way [60].

Since the third period (2011–2020) was the most prolific for this field of research, it is analyzed in detail below. To achieve this, Figure 7 was created, which shows a strategic diagram of a two-dimensional space that was built by plotting the themes according to their centralities and their density rank values. This includes four quadrants, each containing a specific theme: (i) Motor themes, in the upper-right quadrant; (ii) Basic and transversal themes, in the lower-right quadrant; (iii) Highly developed and isolated themes, in the upper-left quadrant; and (iv) Emerging or declining themes, in the lower-left quadrant [61].

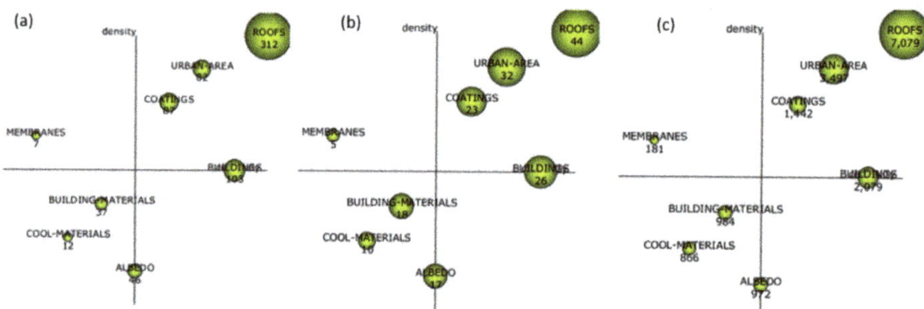

Figure 7. Strategic diagram of the third period, the volume of the spheres is proportional to the number of documents published (**a**), to the h-index (**b**) and to the number of citations (**c**) in the third period associated with each theme.

According to Figure 7a, almost 70% of the published records represent the roofs, urban-area, and coatings themes, and they have the highest impacts (Figure 7b). It can be seen that roofs and urban-area are the most cited themes, with 7079 and 3497 citations, respectively, followed at a distance by coatings, with 1442 citations (Figure 7c). In the case of the roofs theme, its origins show an association with the concepts, building, solar-radiation, solar energy, and urban areas (Figure 6), and it represents the most applied domain of cool materials, as is shown in Figure 4. It is noteworthy that the urban-area motor theme is associated with all of the specific themes of the third period as an origin theme.

The most detailed analysis of the cluster network allowed for the obtainment of information regarding the materials applied for cool surface purposes. Thus, the coatings theme showed a strong relation with the development of pigments (Figure 6a), especially the ones performing in the NIR spectrum as reflective materials, as can be seen in its cluster network (Figure 8d). In terms of the isolated membranes theme (Figure 8c), it mainly concerns the latent strategy of the thermal energy storage resumed in the PCMs. Finally, the two emerging themes of building-materials and cool-materials (Figure 8a,d, respectively) discuss ceramic materials and glass, and retroreflective materials (RR materials), respectively. The most relevant contributions to these topics are analyzed in the next section for the review of materials applied for cool surfaces.

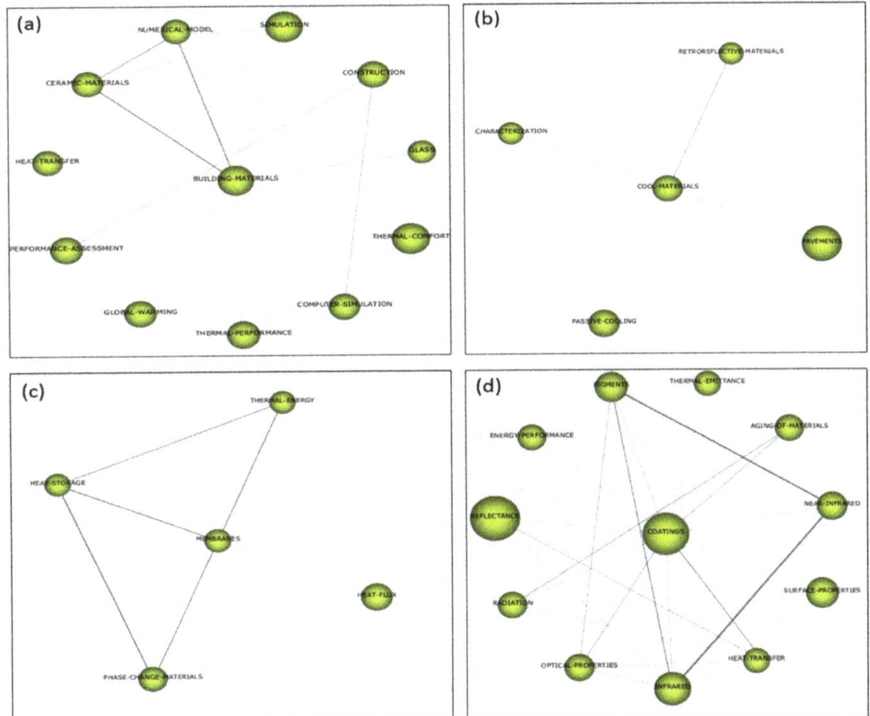

Figure 8. Cluster networks of: (**a**) building-materials; (**b**) cool-materials; (**c**) membranes; and (**d**) coatings.

3.3. *Literature Revision of Materials Applied for Cool Surfaces*

The use of cool materials for building envelopes and pavements helps to increase the solar reflectance index, which decreases the energy use for the cooling demand and enhances the indoor and outdoor thermal comfort. The fabrication of such materials to mitigate the UHI implies the creation of an equilibrium between the technical use of these materials and their environmental impact in order to secure a sustainable production and service loop. In this section, the most applied materials are analyzed: (i) Pigments; (ii) RR materials; (ii) PCMs; (iv) Ceramic materials; and (v) Glass. The implementation of more than one strategy to create a cool surface could be complementary.

These materials were the subject of 63 publications, from which ~25% evaluate ceramic materials, 22% are dedicated to pigments, and 19, 19, and 14% evaluate PCMs, glass, and RR materials, respectively. This sample was selected by means of clustering during the third period. However, six publications were extracted from the previous periods and added to the analysis because of the impacts and contributions of the authors, such as Akbari, H, who has made the most contributions in terms of the records published related to the field of cool surfaces. The literature reviewing of these materials is included below.

3.3.1. Pigments

According to the publications presented in Table 1, the pigments were evaluated through the 14 most relevant publications that are related to the theme of energy use in buildings. Most of the publications are affiliated with research entities based in India and China.

Table 1. Most important publications about pigments.

Title of Publ.	Authors	Journal	Year of Publ.	Ref.	Country of Affiliation
Solar spectral optical properties of pigments—Part I: Model for deriving scattering and absorption coefficients from transmittance and reflectance measurements.	Levinson, R., Berdahl, P., and Akbari, H.	*Solar Energy Materials and Solar Cells*	2005	[27]	United States
Solar spectral optical properties of pigments—Part II: Survey of common colorants.	Levinson, R., Berdahl, P., and Akbari, H.	*Solar Energy Materials and Solar Cells*	2005	[28]	United States
Bismuth titanate as an infrared reflective pigment for cool roof coating.	Meenakshi, P., and Selvaraj, M.	*Solar Energy Materials and Solar Cells*	2018	[29]	India
Pigmentary colors from yellow to red in $Bi_2Ce_2O_7$ by rare earth ion substitutions as possible high NIR reflecting pigments.	Raj, A. K. V., Rao, P. P., Sreena, T. S., and Thara, T. R. A.	*Dyes and Pigments*	2019	[30]	India
A new member of solar heat-reflective pigments: $BaTiO_3$ and its effect on the cooling properties of ASA (acrylonitrile-styrene-acrylate copolymer).	Xiang, B., and Zhang, J.	*Solar Energy Materials and Solar Cells*	2018	[31]	China
Terbium doped Sr_2MO_4 [M = Sn and Zr] yellow pigments with high infrared reflectance for energy saving applications.	Raj, A. K. V., PrabhakarRao, P., Divya, S., and Ajuthara, T. R.	*Powder Technology*	2017	[32]	India
Pigments based on terbium-doped yttrium cerate with high NIR reflectance for cool roof and surface coating applications.	Raj, A. K. V., Prabhakar Rao, P., Sameera, S., and Divya, S.	*Dyes and Pigments*	2015	[33]	India
Study of solar heat-reflective pigments in cool roof coatings.	Cheng, M., Ji, J., and Chang, Y.	*Journal of Beijing University of Chemical*	2009	[62]	China
Surfactant effect on titanium dioxide photosensitized oxidation of 4-dodecyloxybenzyl alcohol.	Bettoni, M., Brinchi, L., Del Giacco, T., Germani, R., Meniconi, S., Rol, C., and Sebastiani, G. V.	*Journal of Photochemistry and Photobiology A: Chemistry*	2012	[63]	Italy
On a cool coating for roof clay tiles: Development of the prototype and thermal-energy assessment.	Pisello, A. L., Cotana, F., and Brinchi, L.	*Energy Procedia*	2014	[64]	Italy
Effects of added ZnO on the crystallization and solar reflectance of titanium-based glaze.	Li, Z., Yang, Y., Peng, C., and Wu, J.	*Ceramics International*	2017	[65]	China
Performance of near-infrared reflective tile roofs.	Thongkanluang, T., Wutisatwongkul, J., Chirakanphaisarn, N., and Pokaipisit, A.	*Advanced Materials Research*	2013	[66]	Thailand
Environmental impact of cool roof paint: case study of house retrofit in two hot islands.	Emmanuel, S., Valentina, S., Petra, G., and Maria, K.	*Energy and Buildings*	2020	[67]	United Kingdom
Preparation of phthalocyanine blue/rutile TiO_2 composite pigment with a ball milling method and study on its NIR reflectivity	Lingyun, C., Xuening, F., Hongbin, Z., and Changliang, H.	*Dyes and Pigments*	2020	[68]	China

Pigments were proven to enhance the NIR reflectance for cool coatings and to maintain a wider range of color choices, which opened the way for an industry of geoengineering and chemical solutions for cool materials [27,28,31,33,62]. The absorption coefficient and the solar spectral backscattering measured in the range of 300–2500nm determines whether the pigment should be implemented in a cool coating [27]. According to Levinson et al. [28], a pigment with a low absorptance is considered cool, whereas a pigment with high NIR transmittance will necessitate an NIR-reflective background (usually white or metallic) in order to form an NIR-reflecting coating. The developed reflective pigments allowed for the transition from lighter colors to dark coating materials, often without compromising the reflective performance. Five dark-colored pigments were analyzed in a study to develop new species of high-NIR reflectance. With a formulation of a 25% weight content of rutile-type titanium dioxide, the white ceramic microspheres, with a 13% weight content, and the heavy calcium carbonate, with a 10% weight content, obtained positive results, and the back temperatures were lowered by 10–20 °C, compared to normal coatings [62].

In general, the pigments integrated into the matrix of cool materials range from organic, of which most are considered transparent, to complex inorganic color mixed-metal oxides that are often opaque [27,28]. Cool pigments achieve a high solar reflectance of up to 95%, compared to TiO_2, such as bismuth titanate [29,30], barium titanate [31], and terbium-doped yttrium cerate [32,33]. The photocatalytically active white inorganic pigment TiO_2 represents the most widely used nontoxic pigment and is characterized by strong scattering, weak absorption, and good stability [63]. As a result, most studies have been compared to its reflective performance [64]. The use of developed pigments was recently the most processed strategy. When incorporated into glazes, fillers, and engobes, synthesized pigments are used in coatings in order to enhance the optical and thermal performances of cool materials, as well as to respect the aesthetic requirements of a design. Several oxides were demonstrated to be able to enhance the NIR reflectance of glazes for concrete cement substrates, steel substrates, metal panels, and clay tiles [33,65,66]. For traditional buildings, using pigments based on sodium silicate for the tile coating improved the solar reflection in the NIR spectrum by 13% without affecting the visual appearance [64]. The integration of such pigments lowers roof overheating and reduces the energy required for cooling.

Cool pigments have now become a growing trend that is taking over cool surface technology, as nanoparticles are integrated into material structures to fulfil a specific design criterion and to enhance the reflective performance. In addition to their chemical stabilities, their effective thermal and optical performances have been proven in numerous studies. However, the acquisition, the production process, the synthesis methods, the high cost of the rare-earth elements used in their composition, and the hazardous environmental effect of the heavy metals may have a negative environmental impact [67,68]. On the other hand, the development of new compositions of reflective pigments continues to increase, reaching more potential evaluations. In addition to the high reflectivity, the research field has also been evaluating the functional and long-term performances against surface contamination [69], thermal insulation [70,71], and the synthesis methods in terms of the energy and raw material costs, which were directed for the sol–gel method and rare-earth compounds [72–77].

3.3.2. RR Materials

According to the publications presented in Table 2, the RR materials were evaluated through the nine most relevant publications that are related to the themes of sustainability, energy use, and construction engineering. Table 2 shows that most of these publications are affiliated with research entities based in Italy.

Table 2. Most important publications about RR materials.

Title of Publ.	Authors	Journal	Year of Publ.	Ref.	Country of Affiliation
Analysis of retro-reflective surfaces for urban heat island mitigation: A new analytical model.	Rossi, F., Pisello, A. L., Nicolini, A., Filipponi, M., and Palombo, M.	*Applied Energy*	2014	[34]	Italy
Experimental evaluation of urban heat island mitigation potential of retro-reflective pavement in urban canyons.	Rossi, F., Castellani, B., Presciutti, A., Morini, E., Anderini, E., Filipponi, M., and Nicolini, A.	*Energy and Buildings*	2016	[35]	Italy
Development and evaluation of directional retroreflective materials: Directional retroreflective materials as a heat island countermeasure.	Sakai, H., Jyota, H., Emura, K., and Igawa, N.	*Journal of Structural and Construction Engineering*	2011	[36]	Japan
A normalization procedure to compare retro-reflective and traditional diffusive materials in terms of UHI mitigation potential.	Gambelli, A. M., Cardinali, M., Filipponi, M., Castellani, B., Nicolini, A., and Rossi, F.	*AIP Conference Proceedings*	2019	[78]	Italy
Reduction of reflected heat by retroreflective materials.	Sakai, H., Emura, K., and Igawa, N.	*Journal of Structural and Construction Engineering*	2008	[79]	Japan
Design, characterization, and fabrication of solar-retroreflective cool-wall materials.	Ronnen, L., Sharon, C., Jonathan, S., Howdy, G., Tatsuya, H., Paul, B.	*Solar Energy Materials and Solar Cells*	2020	[80]	United States, Japan
Optimized retro-reflective tiles for exterior building element.	Morini, E., Castellani, B., Anderini, E., Presciutti, A., Nicolini, A., and Rossi, F.	*Sustainable Cities and Society*	2018	[81]	Italy
Retroreflective façades for urban heat island mitigation: Experimental investigation and energy evaluations.	Rossi, F., Castellani, B., Presciutti, A., Morini, E., Filipponi, M., Nicolini, A., and Santamouris, M.	*Applied Energy*	2015	[82]	Italy, Greece
Optic-energy and visual comfort analysis of retro-reflective building plasters.	B., Castellani, Alberto, M., Andrea, N., Federico, R.	*Building and Environment*	2020	[83]	Italy

Independently of the incidence direction, retroreflectivity refers to the capacity to reflect the incoming light beam to a surface back towards its source [34,35]. The application of diffusive materials on building envelopes induces multiple reflections within the urban canyon patterns; therefore, in order to reduce the captured solar radiation energy, the use of RR materials presents a good alternative [36]. In this sense, RR materials were studied to evaluate their potential with respect to diffusive (Lambertian) coatings, which allowed for the determination of a corrective parameter to enhance the comparison in terms of mitigating the UHI effect [78]. For the retroreflectivity measurements, Sakai et al. [79] present a procedure to measure only the retroreflective components of RR materials, which consists of: (i) Measuring the total reflectance by thermal measurements; (ii) Then measuring the reflectance without retroreflection using a spectrometer; and finally (iii) The RR components are measured by subtracting the latter from the former. Several studies have proved the efficiency of RR materials to reduce the heat trapped in the building surroundings [34–36].

An RR facade with an albedo of 0.60 could reflect 55% of the incident sunlight, whereas a diffusive facade could reflect only 36% with the same albedo [80]. RR materials were also applied in pavement, where the cooling potential could reach a maximum of a 4.6% albedo increase, compared to the traditional white and beige diffusive cool materials [35]. The RR performance of glass beads was discussed in several investigations that show promising results; however, when comparing a base ceramic tile coated with glass spheres and clear solid barium titanate spheres, the latter had the highest global reflectance (39%), and a radiation energy that reflected up to 5% [81].

Besides the advantages that RR materials offer in road-sign use and visual technology, their application in building facades and pavements alleviates the heat trapped inside the buildings that is created by diffuse reflective materials. However, their retroreflective behavior is limited to low angles of incidence, whereas, for high angles of incidence, the solar radiation is symmetrically reflected with regard to the perpendicular radiation [82], which limits the performance of RR materials at all angles of incidence [80,83].

Recently, more research has been oriented toward the performance of RR materials for different angles of incidence that takes into account the irradiated surface geometry scenarios, the urban density, the microclimate, the durability, and the costs and benefits [84]. As a solution, an angular selective behavior was discussed to overcome the limitations of RR materials, especially in summer [85].

3.3.3. PCMs

According to the publications presented in Table 3, the PCM materials were evaluated through the 12 most relevant publications that are related to the theme of energy use in buildings. Most of the publications are affiliated with research entities based in South Korea and the United States.

Table 3. Most important publications about PCMs.

Title of Publ.	Authors	Journal	Year of Publ.	Ref.	Country of Affiliation
Simulating the effects of cool roof and PCM (phase change materials) based roof to mitigate UHI (urban heat island) in prominent US cities.	Roman, K. K., O'Brien, T., Alvey, J. B., and Woo, O. J.	*Energy*	2016	[16]	United States
Prefabricated building units and modern methods of construction (MMC).	Mapston, M., and Westbrook, C.	*Materials for Energy Efficiency and Thermal Comfort in Buildings*	2010	[37]	United Kingdom
Understanding a potential for application of phase-change materials (PCMs) in building envelopes.	Kośny, J., and Kossecka, E.	*ASHRAE Transactions*	2013	[38]	Poland, United States
Thermal stress reduction in cool roof membranes using phase change materials (PCM).	Saffari, M., Piselli, C., de Gracia, A., Pisello, A. L., Cotana, F., and Cabeza, L. F.	*Energy and Buildings*	2018	[39]	Italy, Spain
Development of PCM cool roof system to control urban heat island considering temperate climatic conditions.	Chung, M. H., and Park, J. C.	*Energy and Buildings*	2016	[40]	South Korea
PCM cool roof systems for mitigating urban heat island—an experimental and numerical analysis.	Yang, Y. K., Kim, M. Y., Chung, M. H., and Park, J. C.	*Energy and Buildings*	2019	[43]	South Korea

Table 3. *Cont.*

Title of Publ.	Authors	Journal	Year of Publ.	Ref.	Country of Affiliation
Thermal Performance Test of a Phase-Change-Material Cool Roof System by a Scaled Model.	Yoon, S. G., Yang, Y. K., Kim, T. W., Chung, M. H., and Park, J. C.	*Advances in Civil Engineering*	2018	[86]	Republic of Korea
Numerical analysis of phase change materials/wood-plastic composite roof module system for improving thermal performance.	Seong, J., Seunghwan, W., Hyun, M., Su-Gwang, J., Sumin, K.	*Journal of Industrial and Engineering Chemistry*	2020	[87]	United States, Republic Korea
How to enhance thermal energy storage effect of PCM in roofs with varying solar reflectance: Experimental and numerical assessment of a new roof system for passive cooling in different climate conditions.	Piselli, C., Castaldo, V. L., and Pisello, A. L.	*Solar Energy*	2019	[88]	Italy
Effects of accelerated weathering on the optical characteristics of reflective coatings for cool pavement.	Ning, X., Hui, L., Hengji, Z., Xue, Z., Ming, J.	*Solar Energy Materials and Solar Cells*	2020	[89]	China
Phase change materials for pavement applications: A review	B.R. Anupam, Umesh Chandra Sahoo, PrasenjitRath	*Construction and Building Materials*	2020	[90]	India
Review of current state of research on energy storage, toxicity, health hazards and commercialization of phase changing materials.	S.S.Chandel, Tanya A.	*Renewable and Sustainable Energy Reviews*	2020	[91]	India

PCMs have the ability to change their physical characteristics during phase transition [37]. To compensate for the possible heating load increase in winter while using cool roofs, these materials can prevent the overheating of the roof surface during the summer without increasing the heating load in the winter [86]. As a consequence, they decrease the thermal stress and the annual energy load consumption, in addition to providing thermal inertia for buildings when the melting temperature is optimized [39]. The building energy performance and the thermal comfort could be improved depending on the phase change temperature adopted, according to Chang et al. [87]. Better results were registered for 30 °C than for 20 °C.

The performance of PCM-based surface technology as a UHI mitigation strategy has been evaluated through several studies. For roof application, a cool polyurethane-based membrane reduces the roof-surface temperature and the heat flux through the roof more than a traditional dark bitumen membrane; moreover, the outdoor environmental conditions and the type of PCM host material could influence the performance of the membrane [88]. Similar results were proven using PCM-doped tiles for a cool roof system in simulated summer conditions [40]. In real winter conditions, the use of PCMs maintained a higher indoor temperature than cool paints, and with a low surface temperature reducing the heat penalty [43]. In this sense, several types of PCMs have been tested as dynamic components in buildings in comparison to conventional cool roof materials, showing that their incorporation in the roof materials matrix decreased the heat gain flux by 54%, and registered for various values of albedo compared to cool roof technology, and a lower sensible heat by 40% [16].

PCMs have also been tested for cool pavements. Their use is based on the temperature regulation performance of these materials, which use a lightweight aggregate with a reasonable gradation for better results. The composite PCMs incorporated into the asphalt mixtures achieved a satisfactory cooling performance; however, some of them minimized the strength reductions of the mixtures [89].

The incorporation of PCMs in cool surfaces is emerging as a growing field of research because of their ability to restore energy with a minimum change in volume, and without an increase in temperature; however, the encapsulation method may affect their performance and cause leakage [16,38–40,88,90]. Moreover, some PCMs could be classified as "unsustainable"; for instance, the PCM paraffin wax that is commonly used releases toxic vapors when burnt, which can result in severe health hazards, as it contains formaldehyde, benzene, toluene, and other toxic compounds [91].

In order to enhance the thermal energy storage in buildings, recent investigations have been oriented toward developing new techniques of encapsulation. Nano/microencapsulation methods have presented promising results while avoiding leakage [92,93]. Moreover, different combinations of reflective coatings and PCM applications were tested; the optimal combination required a layer of thermal insulation between the two materials, which incorporates the PCMs into the buildings and is a complementary technique for all the energy-saving system [94].

3.3.4. Ceramic Materials

According to the publications presented in Table 4, ceramic materials have been evaluated in 16 publications that are highly related to the theme of energy use in buildings. Nearly 50% of the publications are affiliated, either implicitly or explicitly, with research entities based in Italy.

Table 4. Most important publications about ceramic materials.

Title of Publ.	Authors	Journal	Year of Publ.	Ref.	Country of Affiliation
Experimental evaluation of thermal performance of cool pavement material using waste tiles in tropical climate.	Anting, N., Md. Din, M. F., Iwao, K., Ponraj, M., Jungan, K., Yong, L. Y., and Siang, A. J. L. M.	*Energy and Buildings*	2017	[44]	Malaysia, Japan
On a cool coating for roof clay tiles: Development of the prototype and thermal-energy assessment.	Pisello, A. L., Cotana, F., and Brinchi, L.	*Energy Procedia*	2014	[64]	Italy
Performance of near-infrared reflective tile roofs.	Thongkanluang, T., Wutisatwongkul, J., Chirakanphaisarn, N., and Pokaipisit, A.	*Advanced Materials Research*	2013	[66]	Thailand
Design of ceramic tiles with high solar reflectance through the development of a functional engobe.	Ferrari, Chiara, Libbra, A., Muscio, A., and Siligardi, C.	*Ceramics International*	2013	[95]	Italy
Design of a cool color glaze for solar reflective tile application.	Ferrari, C., Muscio, A., Siligardi, C., and Manfredini, T.	*Ceramics International*	2015	[96]	Italy
High-solar-reflectance building ceramic tiles based on titanite (CaTiSiO5) glaze.	Li, Z., Zhao, M., Zeng, J., Peng, C., and Wu, J.	*Solar Energy*	2017	[97]	China
Cooler tile-roofed buildings with near-infrared-reflective non-white coatings.	Levinson, R., Akbari, H., and Reilly, J. C.	*Building and Environment*	2007	[98]	United States
Measured temperature reductions and energy savings from a cool tile roof on a central California home.	Rosado, P. J., Faulkner, D., Sullivan, D. P., and Levinson, R.	*Energy and Buildings*	2014	[99]	United States

Table 4. *Cont.*

Title of Publ.	Authors	Journal	Year of Publ.	Ref.	Country of Affiliation
New strategy to mitigate urban heat island effect: Energy saving by combining high albedo and low thermal diffusivity in glass ceramic materials.	Enríquez, E., Fuertes, V., Cabrera, M. J., Seores, J., Muñoz, D., and Fernández, J. F.	*Solar Energy*	2017	[100]	Spain
White sintered glass-ceramic tiles with improved thermal insulation properties for building applications.	Marangoni, M., Nait-Ali, B., Smith, D. S., Binhussain, M., Colombo, P., and Bernardo, E.	*Journal of the European Ceramic Society*	2017	[101]	Italy, France, Saudi Arabia, United States
A composite cool colored tile for sloped roofs with high "equivalent" solar reflectance.	Ferrari, Chiara, Libbra, A., Cernuschi, F. M., De Maria, L., Marchionna, S., Barozzi, M., . . . Muscio, A.	*Energy and Buildings*	2016	[102]	Italy
Optical properties of traditional clay tiles for ventilated roofs and implication on roof thermal performance.	Di Giuseppe, E., Sabbatini, S., Cozzolino, N., Stipa, P., and D'Orazio, M.	*Journal of Building Physics*	2019	[103]	Italy
Study on the cool roof effect of Japanese traditional tiled roof: Numerical analysis of solar reflectance of unevenness tiled surface and heat budget of typical tiled roof system.	Takebayashi, H., Moriyama, M., and Sugihara, T.	*Energy and Buildings*	2012	[104]	Japan
Development of clay tile coatings for steep-sloped cool roofs.	Pisello, A. L., Cotana, F., Nicolini, A., and Brinchi, L.	*Energies*	2013	[105]	Italy
Thermal-energy analysis of roof cool clay tiles for application in historic buildings and cities.	Pisello, A. L.	*Sustainable Cities and Society*	2015	[106]	Italy
Study on roof tile's colors in Malaysia for development of new anti-warming roof tiles with higher Solar Reflectance Index (SRI).	Yacouby, A. M. A., Khamidi, M. F., Nuruddin, M. F., Farhan, S. A., and Razali, A. E.	*National Postgraduate Conference—Energy and Sustainability: Exploring the Innovative Minds*	2011	[107]	Malaysia

The publications showed the development of innovative solutions for a better solar reflectance index of ceramic materials, such as tiles, glazes, and engobes. The use of ceramic tiles is considered an effective component of the cool roof strategy, thanks to its durability and its solar properties, especially if it is glazed [95–97]. The substrate material was tested with the application of different ceramic coatings through several studies, and the developed nonwhite coatings enhanced the solar reflective performance and showed interesting results in terms of energy saving [98,99]. Moreover, the wollastonite–hardystonite glass–ceramic porous tiles showed high reflectances of solar radiation, coupled with low thermal conductivity in an arid environment [100], which highlights the complementary function with regard to the thermal insulation properties [101,102]. In the same sense, an improvement in the thermal performance of a residential building was found during the summer and the winter, and 75% of the solar radiation reflectance in the NIR spectrum was registered, i.e., 10% more with respect to traditional tiles, without altering the visible appearance [64]. The latter property is highly considered for historical buildings that are required to maintain their original aesthetic appearance. These types of buildings often exist in the center of urban areas, which are strongly affected by the UHI. Their retrofitting using innovative cool clay tile coatings increased the solar reflectance by 20% while maintaining the original color intact. This kind of retrofitting enhances the thermal responses of buildings and the urban climate in general [103–106].

Glazes present a good complementary component for cool tiles as glass–ceramic materials; they are fabricated through a controlled crystallization process for a desired microstructure. The incorporation of cool pigments in the composition of glazes yields the total reflective performance of the product up to 82.8% [66]. The results of an experimental study show that a tile coated with glass ceramic material induces 20% energy savings, compared to TiO_2-based paints, and that it could be used for roofs and pavements [107]. The application of engobes enhances the solar reflectance as well, by up to 0.90 [95]. As an intermediate layer between the substrate and the glaze, it provides a high degree of adhesion while taking into account the convenience of all the coefficients of thermal expansion.

The application of developed glazes and engobes enhances the solar reflective performance of tiles [108]. In recent studies, the development of cool ceramic coatings was strongly discussed in terms of the low-cost routes, the use of secondary materials, and the self-cleaning abilities [109]. In turn, research has been focused on enhancing the compositions of glass ceramic frits, opacifiers, and pigments to reach the optimal potential for NIR reflectance. The development of coatings has been oriented toward the use of dynamic coatings, such as the passive ones: photochromic and thermochromic coatings [110]. It is clear that the research field is an intersection of multiple complementary techniques and materials, which include pigments, glazes, and tiles [73,111], and this creates a wide range of opportunities to attain the energy-saving potential of cool materials in future studies.

3.3.5. Glass

According to the publications presented in Table 5, the use of glass is discussed through the 12 most relevant publications that are related to the themes of buildings and the cleaner production of materials. The publications are affiliated with research entities based in different countries and that are not concentrated in a specific one.

Table 5. Most important publications about glass.

Title of Publ.	Authors	Journal	Year of Publ.	Ref.	Country of Affiliation
Optic-energy and visual comfort analysis of retro-reflective building plasters.	Castellani, Gambelli, Nicolini, Rossi	*Building and Environment*	2020	[83]	Italy
Waste glass in civil engineering applications—A review.	Kazmi, D., Williams, D. J. and Serati, M.	*International Journal of Applied Ceramic Technology*	2020	[112]	Australia
Reuse of waste glass in building brick production.	Demir, I.	*Waste Management and Research*	2009	[113]	Turkey
Utilization of waste glass to enhance physical-mechanical properties of fired clay brick.	Phonphuak, N., Kanyakam, S., and Chindaprasirt, P.	*Journal of Cleaner Production*	2016	[114]	Thailand
Properties of Fired Clay Bricks Mixed with Waste Glass.	Abdeen, H., and Shihada, S.	*Journal of Scientific Research and Reports*	2017	[115]	Palestine
The role of glass waste in the production of ceramic-based products and other applications: A review.	Silva, R. V., de Brito, J., Lye, C. Q., and Dhir, R. K.	*Journal of Cleaner Production*	2017	[116]	Portugal, United Kingdom
Effect of waste glass on properties of burnt clay bricks.	Hameed, A., Haider, U., Qazi, A. U., and Abbas, S.	*Pakistan Journal of Engineering and Applied Sciences*	2018	[117]	Canada
Thermal performance evaluation of eco-friendly bricks incorporating waste glass sludge.	Kazmi, S. M. S., Munir, M. J., Wu, Y. F., Hanif, A., and Patnaikuni, I.	*Journal of Cleaner Production*	2018	[118]	Australia, Pakistan, Hong Gong

Table 5. *Cont.*

Title of Publ.	Authors	Journal	Year of Publ.	Ref.	Country of Affiliation
Glass recycling in the production of low-temperature stoneware tiles.	Lassinantti Gualtieri, M., Mugoni, C., Guandalini, S., Cattini, A., Mazzini, D., Alboni, C., and Siligardi, C.	*Journal of Cleaner Production*	2018	[119]	Italy
Effect of glass powder on the technological properties and microstructure of clay mixture for porcelain stoneware tiles manufacture.	Njindam, O. R., Njoya, D., Mache, J. R., Mouafon, M., Messan, A., and Njopwouo, D.	*Construction and Building Materials*	2018	[120]	Burkina Fasu, Cameroon
Incorporating hollow glass microsphere to cool asphalt pavement: Preliminary evaluation of asphalt mastic.	Du Yinfei, Dai Mingxin, Deng Haibin, Deng Deyi, Cheng Peifeng, Ma Cong	*Construction and Building Materials*	2020	[121]	China
Cool White Polymer Coatings based on Glass Bubbles for Buildings.	Nie, YoungjaeYoo, Hasitha Hewakuruppu, Sullivan, Krishna, Jaeho Lee	*Scientific Reports*	2020	[122]	South Korea, United States

Recently, numerous solutions for cool materials have been developed through the academic research in terms of sustainability, which have paved the way for substituting raw materials and using secondary ones. Waste glass was introduced as an alternative as a solution to the increased need for efficient waste management strategies.

Using glass as a secondary material in the fabrication of ceramic materials enhances the mechanical behavior and the sintering action because of its amorphous structure [112]. Different percentages of waste glass have been tested in different studies for the fabrication of clay bricks, and, in general, the mechanical and thermal behaviors of these materials were affected by the percentage of the substitution, the size of the waste glass particles, and the chemical composition of each type [113–118]. For the fabrication of tiles, 41 wt.% of waste glass demonstrated good flexural strength and abrasion resistance when using the boron-rich waste glass as a sintering promoter [119]. However, Njindam et al. [120] demonstrated that the addition of high amounts of glass (>30 wt.%) into ceramic bodies is undesirable because of its negative effect on the physical properties. The integration of glass into the matrix of cool materials was mostly in its finest structure; in general, good results were obtained when using small-sized particles of glass. For instance, hollow glass microspheres were integrated into an asphalt mixture, which resulted in a 40% decrease in the thermal conductivity, and a 60% increase in the infrared reflectance [121,122]. In the same sense, it was demonstrated that glass spheres incorporated in RR materials showed good results in terms of the energy reflected beyond the canyon for the building envelopes, in addition to the road-traffic-marking efficiency [83].

According to the different studies, each material manifests specific thermal, physical, and mechanical properties, depending on different conditions, such as the chemical composition and the particle size of the waste glass.

The evaluation of the solar reflectance of ceramic materials, such as tiles and bricks, for building envelopes incorporating waste glass has been little discussed in the academic research. Nevertheless, waste glass cullet was successfully mixed with conventional materials to fabricate sustainable asphalt roof shingles, which have a solar reflectance greater than 25%, and better solar reflectance properties [45]. Recent research has shown that the use of waste glass tile coatings should be considered for a global solar reflectance analysis [123]. Future studies are needed to evaluate the influence of using waste glass on the solar radiation reflectance in ceramic materials, especially in the NIR solar spectrum, where 52% of the solar energy is concentrated.

4. Conclusions

On the basis of a bibliometric analysis using SciMAT software, a sample of 982 academic records in the field of cool surfaces was processed for three periods, from 1995 to 2020. The most prolific period, between 2011 and 2020, with 87.74% of the total records, was analyzed specifically. It was shown that cool surfaces are a developing research field that aims to create sustainable cool materials and coatings, with efficient solar reflectivity, that conform to the retrofitting requirements, and that are adaptable to the specificities of each domain of application: pavements, facades, and particularly roofs.

The research field has placed emphasis on the materials perspective, as most materials trending in this topic are pigments, RR materials, PCMs, ceramic materials, and glass. The heat energy mostly falls within the NIR wavelength region; thus, several recent studies have been developing solar radiation reflective materials. However, the efficiency and the manufacturing of these materials depend not only on the good thermal and optical performances, but also on the environmental impact during the production and the service life. The results of the analysis of the key publications are summarized below:

- Developing pigments with high NIR radiation reflectance is a growing industrial domain, which provides up to 95% of the solar reflectivity performance. However, the acquisition, the production process, and the methods of synthesis of these nanoparticles may inconveniently affect the environment;
- Retroreflective materials, such as backscattering materials, present good performances by reflecting the incident solar radiation beyond the urban canyon and easing the heat trapped in the urban canopy with respect to the diffusive materials. However, their retroreflective behavior occurs mainly for low angles of incidence, therefore limiting their performance at large angles of incidence. More studies should discuss alternative solutions for the angular selective behavior;
- PCMs represent a good solution as cool-surface dynamic switch materials, and their implementation in the reflective materials matrix enhances the solar reflectivity performance. However, some PCM-based materials may contain toxic metals, and the encapsulation methods should be further addressed in future studies for better potential uses;
- Ceramic materials are complementary effective materials for cool materials and coatings that enhance the total solar reflectance performance, with a broad range of reflective glazes and engobes;
- Most of the studies discuss the integration of glass in the fabrication of ceramic materials as fine particles. The use of glass in these materials, with the optimal dosage and particle size, enhances their physical, mechanical, and thermal properties;
- All the cool materials discussed in this review have a common aim, and that is to conform to energy efficiency while using sustainable materials and methods, these materials could be used as complementary strategies for an energy saving system.
- The use of secondary materials as substitutes for the manufacturing of cool materials have paved the way toward more sustainable solutions. The incorporation of glass generated from waste in the manufacturing of cool materials is poorly discussed in the academic literature; thus, more research on the solar reflectance performance of waste glass particles as raw material should be further investigated. This study presents a scientific contribution to the cool surfaces research field in terms of the materials and strategies it presents to counter the urban heat island effect.

Author Contributions: Conceptualization, experiments, writing, formal analysis, and the postprocessing, C.M. and M.M.-M.; conceptualization, writing, project administration, funding acquisition, and supervision, M.Z. and D.P.R. All authors have read and agreed to the published version of the manuscript.

Funding: This research was funded by the State Research Agency (AEI) of Spain and the European Regional Development Funds (ERDF), under project: PID2019-108761RB-I00.

Institutional Review Board Statement: Not applicable.

Informed Consent Statement: Not applicable.

Data Availability Statement: Data are available upon request.

Acknowledgments: The research group TEP-968 technologies for the circular economy of the University of Granada, Spain.

Conflicts of Interest: The authors declare no conflict of interest.

References

1. Gago, E.J.; Roldan, J.; Pacheco-Torres, R.; Ordonez, J. The city and urban heat islands: A review of strategies to mitigate adverse effects. *Renew. Sustain. Energy Rev.* **2013**, *25*, 749–758. [CrossRef]
2. Rosenfeld, A.H.; Akbari, H.; Bretz, S.; Fishman, B.L.; Stamper-Kurn, D.; Sailor, D.; Taha, H. Mitigation of urban heat islands: Materials, utility programs, updates. *Energy Build.* **1995**, *22*, 255–265. [CrossRef]
3. Zinzi, M.; Agnoli, S. Cool and green roofs. An energy and comfort comparison between passive cooling and mitigation urban heat island techniques for residential buildings in the Mediterranean region. *Energy Build.* **2012**, *55*, 66–76. [CrossRef]
4. Mohajerani, A.; Bakaric, J.; Jeffrey-Bailey, T. The urban heat island effect, its causes, and mitigation, with reference to the thermal properties of asphalt concrete. *J. Environ. Manag.* **2017**, *197*, 522–538. [CrossRef]
5. Akbari, H.; Kolokotsa, D. Three decades of urban heat islands and mitigation technologies research. *Energy Build.* **2016**, *133*, 834–842. [CrossRef]
6. Akbari, H.; Davis, S.; Dorsano, S.; Huang, J.; Winnett, S. *Cooling Our Communities: A Guidebook on Tree Planting and Light-Colored Surfacing*; U.S. Environmental Protection Agency: Washington, DC, USA, 1992.
7. Kolokotsa, D.; Santamouris, M.; Zerefos, S. Green and cool roofs' urban heat island mitigation potential in European climates for office buildings under free floating conditions. *Sol. Energy* **2013**, *95*, 118–130. [CrossRef]
8. Chatterjee, S.; Khan, A.; Dinda, A.; Mithun, S.; Khatun, R.; Akbari, H.; Kusaka, H.; Mitra, C.; Bhatti, S.S.; Van Doan, Q.; et al. Simulating micro-scale thermal interactions in different building environments for mitigating urban heat islands. *Sci. Total Environ.* **2019**, *663*, 610–631. [CrossRef]
9. Santamouris, M. Using cool pavements as a mitigation strategy to fight urban heat island—A review of the actual developments. *Renew. Sustain. Energy Rev.* **2013**, *26*, 224–240. [CrossRef]
10. Qin, Y. A review on the development of cool pavements to mitigate urban heat island effect. *Renew. Sustain. Energy Rev.* **2015**, *52*, 445–459. [CrossRef]
11. Akbari, H.; Cartalis, C.; Kolokotsa, D.; Muscio, A.; Pisello, A.L.; Rossi, F.; Santamouris, M.; Synnefa, A.; Wong, N.H.; Zinzi, M. Local climate change and urban heat island mitigation techniques—the state of the art. *J. Civ. Eng. Manag.* **2015**, *22*, 1–16. [CrossRef]
12. Lai, D.; Liu, W.; Gan, T.; Liu, K.; Chen, Q. A review of mitigating strategies to improve the thermal environment and thermal comfort in urban outdoor spaces. *Sci. Total Environ.* **2019**, *661*, 337–353. [CrossRef] [PubMed]
13. Zhu, Z.; Zhou, D.; Wang, Y.; Ma, D.; Meng, X. Assessment of urban surface and canopy cooling strategies in high-rise residential communities. *J. Clean. Prod.* **2020**, *288*, 125599. [CrossRef]
14. Yang, J.; Wang, Z.-H.; Kaloush, K.E. Environmental impacts of reflective materials: Is high albedo a 'silver bullet' for mitigating urban heat island? *Renew. Sustain. Energy Rev.* **2015**, *47*, 830–843. [CrossRef]
15. Hosseini, M.; Akbari, H. Effect of cool roofs on commercial buildings energy use in cold climates. *Energy Build.* **2016**, *114*, 143–155. [CrossRef]
16. Roman, K.K.; O'Brien, T.; Alvey, J.; Woo, O. Simulating the effects of cool roof and PCM (phase change materials) based roof to mitigate UHI (urban heat island) in prominent US cities. *Energy* **2016**, *96*, 103–117. [CrossRef]
17. Pisello, A.L. State of the art on the development of cool coatings for buildings and cities. *Sol. Energy* **2017**, *144*, 660–680. [CrossRef]
18. Akbari, H.; Matthews, D. Global cooling updates: Reflective roofs and pavements. *Energy Build.* **2012**, *55*, 2–6. [CrossRef]
19. Gao, Y.; Xu, J.; Yang, S.; Tang, X.; Zhou, Q.; Ge, J.; Xu, T.; Levinson, R. Cool roofs in China: Policy review, building simulations, and proof-of-concept experiments. *Energy Policy* **2014**, *74*, 190–214. [CrossRef]
20. Lee Shoemaker, W. Cool metal roofs provide long-term solutions. *Constr. Specif.* **2003**, *56*, 64–69.
21. Chen, M.Z.; Wei, W.; Wu, S.P. On Cold Materials of Pavement and High-Temperature Performance of Asphalt Concrete. *Mater. Sci. Forum* **2009**, *620–622*, 379–382. [CrossRef]

22. Smith, G.B.; Labarias, M.A.G.; Arnold, M.D.; Gentle, A.R. Super-cool paints: Optimizing composition with a modified four-flux model. In Proceedings of the Thermal Radiation Management for Energy Applications, San Diego, CA, USA, 6–10 August 2017. [CrossRef]

23. Bretz, S.; Akbari, H.; Rosenfeld, A. Practical issues for using solar-reflective materials to mitigate urban heat islands. *Atmos. Environ.* **1998**, *32*, 95–101. [CrossRef]

24. Muniz-Miranda, F.; Minei, P.; Contiero, L.; Labat, F.; Ciofini, I.; Adamo, C.; Bellina, F.; Pucci, A. Aggregation Effects on Pigment Coatings: Pigment Red 179 as a Case Study. *ACS Omega* **2019**, *4*, 20315–20323. [CrossRef] [PubMed]

25. Rosati, A.; Fedel, M.; Rossi, S. NIR reflective pigments for cool roof applications: A comprehensive review. *J. Clean. Prod.* **2021**, *313*, 127826. [CrossRef]

26. Xie, N.; Li, H.; Abdelhady, A.; Harvey, J. Laboratorial investigation on optical and thermal properties of cool pavement nano-coatings for urban heat island mitigation. *Build. Environ.* **2018**, *147*, 231–240. [CrossRef]

27. Levinson, R.; Berdahl, P.; Akbari, H. Solar spectral optical properties of pigments—Part I: Model for deriving scattering and absorption coefficients from transmittance and reflectance measurements. *Sol. Energy Mater. Sol. Cells* **2005**, *89*, 319–349. [CrossRef]

28. Levinson, R.; Berdahl, P.; Akbari, H. Solar spectral optical properties of pigments—Part II: Survey of common colorants. *Sol. Energy Mater. Sol. Cells* **2005**, *89*, 351–389. [CrossRef]

29. Meenakshi, P.; Selvaraj, M. Bismuth titanate as an infrared reflective pigment for cool roof coating. *Sol. Energy Mater. Sol. Cells* **2018**, *174*, 530–537. [CrossRef]

30. Raj, A.K.; Rao, P.P.; Sreena, T.S.; Thara, T.A. Pigmentary colors from yellow to red in $Bi_2Ce_2O_7$ by rare earth ion substitutions as possible high NIR reflecting pigments. *Dye. Pigment.* **2019**, *160*, 177–187. [CrossRef]

31. Xiang, B.; Zhang, J. A new member of solar heat-reflective pigments: $BaTiO_3$ and its effect on the cooling properties of ASA (acrylonitrile-styrene-acrylate copolymer). *Sol. Energy Mater. Sol. Cells* **2018**, *180*, 67–75. [CrossRef]

32. Raj, A.K.; Rao, P.P.; Divya, S.; Ajuthara, T. Terbium doped Sr_2MO_4 [M = Sn and Zr] yellow pigments with high infrared reflectance for energy saving applications. *Powder Technol.* **2017**, *311*, 52–58. [CrossRef]

33. Raj, A.K.; Rao, P.P.; Sameera, S.; Divya, S. Pigments based on terbium-doped yttrium cerate with high NIR reflectance for cool roof and surface coating applications. *Dye. Pigment.* **2015**, *122*, 116–125. [CrossRef]

34. Rossi, F.; Pisello, A.L.; Nicolini, A.; Filipponi, M.; Palombo, M. Analysis of retro-reflective surfaces for urban heat island mitigation: A new analytical model. *Appl. Energy* **2014**, *114*, 621–631. [CrossRef]

35. Rossi, F.; Castellani, B.; Presciutti, A.; Morini, E.; Anderini, E.; Filipponi, M.; Nicolini, A. Experimental evaluation of urban heat island mitigation potential of retro-reflective pavement in urban canyons. *Energy Build.* **2016**, *126*, 340–352. [CrossRef]

36. Sakai, H.; Iyota, H.; Emura, K.; Igawa, N. Development and evaluation of directional retroreflective materials. *Struct. Constr. Eng. (Trans. AIJ)* **2011**, *76*, 1229–1234. [CrossRef]

37. Mapston, M.; Westbrook, C. Prefabricated building units and modern methods of construction (MMC). *Mater. Energy Effic. Therm. Comf. Build* **2010**, *2010*, 427–454. [CrossRef]

38. Kośny, J.; Kossecka, E. Understanding a potential for application of phase-change materials (PCMs) in building envelopes. *ASHRAE Trans.* **2013**, *119*, 3–13.

39. Saffari, M.; Piselli, C.; de Gracia, A.; Pisello, A.L.; Cotana, F.; Cabeza, L.F. Thermal stress reduction in cool roof membranes using phase change materials (PCM). *Energy Build.* **2018**, *158*, 1097–1105. [CrossRef]

40. Chung, M.H.; Park, J.C. Development of PCM cool roof system to control urban heat island considering temperate climatic conditions. *Energy Build.* **2016**, *116*, 341–348. [CrossRef]

41. Pisello, A.L.; Fortunati, E.; Fabiani, C.; Mattioli, S.; Dominici, F.; Torre, L.; Cabeza, L.F.; Cotana, F. PCM for improving polyurethane-based cool roof membranes durability. *Sol. Energy Mater. Sol. Cells* **2017**, *160*, 34–42. [CrossRef]

42. Yang, Y.K.; Kang, I.S.; Chung, M.H.; Kim, S.; Park, J.C. Effect of PCM cool roof system on the reduction in urban heat island phenomenon. *Build. Environ.* **2017**, *122*, 411–421. [CrossRef]

43. Yang, Y.K.; Kim, M.Y.; Chung, M.H.; Park, J.C. PCM cool roof systems for mitigating urban heat island - an experimental and numerical analysis. *Energy Build.* **2019**, *205*, 109537. [CrossRef]

44. Anting, N.; Din, M.F.M.; Iwao, K.; Ponraj, M.; Jungan, K.; Yong, L.Y.; Siang, A.J.L.M. Experimental evaluation of thermal performance of cool pavement material using waste tiles in tropical climate. *Energy Build.* **2017**, *142*, 211–219. [CrossRef]

45. Kiletico, M.J.; Hassan, M.M.; Mohammad, L.N.; Alvergue, A.J. New Approach to Recycle Glass Cullet in Asphalt Shingles to Alleviate Thermal Loads and Reduce Heat Island Effects. *J. Mater. Civ. Eng.* **2015**, *27*, 04014219. [CrossRef]

46. Cobo, M.J.; López-Herrera, A.G.; Herrera-Viedma, E.; Herrera, F. An approach for detecting, quantifying, and visualizing the evolution of a research field: A practical application to the Fuzzy Sets Theory field. *J. Inform.* **2011**, *5*, 146–166. [CrossRef]

47. Noyons, E.C.M.; Moed, H.F.; Luwel, M. Combining mapping and citation analysis for evaluative bibliometric purposes: A bibliometric study. *J. Am. Soc. Inf. Sci.* **1999**, *50*, 115–131. [CrossRef]

48. Callon, M.; Courtial, J.-P.; Turner, W.A.; Bauin, S. From translations to problematic networks: An introduction to co-word analysis. *Soc. Sci. Inf.* **1983**, *22*, 191–235. [CrossRef]

49. Callon, M.; Courtial, J.P.; Laville, F. Co-word analysis as a tool for describing the network of interactions between basic and technological research: The case of polymer chemsitry. *Scientometrics* **1991**, *22*, 155–205. [CrossRef]

50. Cobo, M.J.; Martínez, M.; Gutiérrez-Salcedo, M.; Fujita, H.; Herrera-Viedma, E. 25years at Knowledge-Based Systems: A bibliometric analysis. *Knowledge-Based Syst.* **2015**, *80*, 3–13. [CrossRef]

51. Zinzi, M.; Fasano, G. Properties and performance of advanced reflective paints to reduce the cooling loads in buildings and mitigate the heat island effect in urban areas. *Int. J. Sustain. Energy* **2009**, *28*, 123–139. [CrossRef]

52. Boriboonsomsin, K.; Reza, F. Mix Design and Benefit Evaluation of High Solar Reflectance Concrete for Pavements. *Transp. Res. Rec. J. Transp. Res. Board* **2007**, *2011*, 11–20. [CrossRef]

53. Berdahl, P.; Bretz, S.E. Preliminary survey of the solar reflectance of cool roofing materials. *Energy Build.* **1997**, *25*, 149–158. [CrossRef]

54. Akbari, H.; Pomerantz, M.; Taha, H. Cool surfaces and shade trees to reduce energy use and improve air quality in urban areas. *Sol. Energy* **2001**, *70*, 295–310. [CrossRef]

55. Rosenfeld, A.H.; Akbari, H.; Romm, J.J.; Pomerantz, M. Cool communities: Strategies for heat island mitigation and smog reduction. *Energy Build.* **1998**, *28*, 51–62. [CrossRef]

56. Zhou, N.; Sha, S.; Zhang, Y.; Li, S.; Xu, S.; Luan, J. Coprecipitation synthesis of a green Co-doped wurtzite structure high near-infrared reflective pigments using ammonia as precipitant. *J. Alloy. Compd.* **2019**, *820*, 153183. [CrossRef]

57. Kavitha, K.; Sivakumar, A. Impact of titanium concentration in structural and optical behaviour of nano $Bi_2Ce_{2-x}Ti_xO_7$ (x = 0–1) high NIR reflective and UV shielding yellow and orange pigments. *Inorg. Chem. Commun.* **2020**, *120*, 108163. [CrossRef]

58. Matias, L.; Gonçalves, L.; Costa, A.; Santos, C.P. Cool Façades - Thermal Performance Assessment Using Infrared Thermography. *Key Eng. Mater.* **2014**, *634*, 14–21. [CrossRef]

59. Cozza, E.S.; Alloisio, M.; Comite, A.; Di Tanna, G.; Vicini, S. NIR-reflecting properties of new paints for energy-efficient buildings. *Sol. Energy* **2015**, *116*, 108–116. [CrossRef]

60. Chen, Y.; Mandal, J.; Li, W.; Smith-Washington, A.; Tsai, C.-C.; Huang, W.; Shrestha, S.; Yu, N.; Han, R.P.S.; Cao, A.; et al. Colored and paintable bilayer coatings with high solar-infrared reflectance for efficient cooling. *Sci. Adv.* **2020**, *6*, eaaz5413. [CrossRef]

61. Cobo, M.J.; López-Herrera, A.G.; Herrera-Viedma, E.; Herrera, F. SciMAT: A new science mapping analysis software tool. *J. Am. Soc. Inf. Sci. Technol.* **2012**, *63*, 1609–1630. [CrossRef]

62. Cheng, M.; Ji, J.; Chang, Y.; Herrera, F. Study of solar heat-reflective pigments in cool roof coatings. *J. Beijing Univ. Chem. Technol. (Nat. Sci. Ed.).* **2009**, *36*, 50–54.

63. Bettoni, M.; Brinchi, L.; Del Giacco, T.; Germani, R.; Meniconi, S.; Rol, C.; Sebastiani, G.V.; Meniconi, S. Surfactant effect on titanium dioxide photosensitized oxidation of 4-dodecyloxybenzyl alcohol. *J. Photochem. Photobiol. A Chem.* **2012**, *229*, 53–59. [CrossRef]

64. Pisello, A.L.; Cotana, F.; Brinchi, L. On a Cool Coating for Roof Clay Tiles: Development of the Prototype and Thermal-energy Assessment. *Energy Procedia* **2014**, *45*, 453–462. [CrossRef]

65. Li, Z.; Yang, Y.; Peng, C.; Wu, J. Effects of added ZnO on the crystallization and solar reflectance of titanium-based glaze. *Ceram. Int.* **2017**, *43*, 6597–6602. [CrossRef]

66. Thongkanluang, T.; Wutisatwongkul, J.; Chirakanphaisarn, N.; Pokaipisit, A. Performance of Near-Infrared Reflective Tile Roofs. *Adv. Mater. Res.* **2013**, *770*, 30–33. [CrossRef]

67. Shittu, E.; Stojceska, V.; Gratton, P.; Kolokotroni, M. Environmental impact of cool roof paint: Case-Study of house retrofit in two hot islands. *Energy Build.* **2020**, *217*, 110007. [CrossRef]

68. Cao, L.; Fei, X.; Zhao, H.; Huang, C. Preparation of phthalocyanine blue/rutile TiO_2 composite pigment with a ball milling method and study on its NIR reflectivity. *Dye. Pigment.* **2019**, *173*, 107879. [CrossRef]

69. Yun, T.H.; Yim, C. Uniform Fabrication of Hollow Titania Using Anionic Modified Acrylated Polymer Template for Phase Composition Effect as Photocatalyst and Infrared Reflective Coating. *Nanomaterials* **2021**, *11*, 2845. [CrossRef]

70. Tian, M.; Chen, C.; Han, A.; Ye, M.; Chen, X. Estimating thermal insulation performance and weather resistance of acrylonitrile-styrene-acrylate modified with high solar reflective pigments: Pr3+/Cr3+ doped BaTiO. *Sol. Energy* **2021**, *225*, 934–941. [CrossRef]

71. Ramos, N.; Maia, J.; Souza, A.; Almeida, R.; Silva, L. Impact of Incorporating NIR Reflective Pigments in Finishing Coatings of ETICS. *Infrastructures* **2021**, *6*, 79. [CrossRef]

72. Soranakom, P.; Vittayakorn, N.; Rakkwamsuk, P.; Supothina, S.; Seeharaj, P. Effect of surfactant concentration on the formation of $Fe_2O_3@SiO_2$ NIR-reflective red pigments. *Ceram. Int.* **2021**, *47*, 13147–13155. [CrossRef]

73. Divya, S.; Das, S. New red pigments based on Li3AlMnO5 for NIR reflective cool coatings. *Ceram. Int.* **2021**, *47*, 30381–30390. [CrossRef]

74. Rosati, A.; Fedel, M.; Rossi, S. Laboratory scale characterization of cool roof paints: Comparison among different artificial radiation sources. *Prog. Org. Coatings* **2021**, *161*, 106464. [CrossRef]

75. Kamal, A.; Abdouss, M.; Mazhar, M. Synthesis, characterization, and visible–near infrared properties of some perylene-3,4,9,10-tetracarboxylic bisimide derivatives. *J. Chem. Technol. Biotechnol.* **2021**, *96*, 2837–2844. [CrossRef]

76. Li, Y.; Ma, Y.; Liu, W.; Wang, Z.; Liu, H.; Wang, X.; Wei, H.; Zeng, S.; Yi, N.; Cheng, G.J. A promising inorganic $YFeO_3$ pigments with high near-infrared reflectance and infrared emission. *Sol. Energy* **2021**, *226*, 180–191. [CrossRef]

77. Zhou, W.; Liu, Y.; Sun, Q.; Ye, J.; Chen, L.; Wang, J.; Li, G.; Lin, H.; Ye, Y.; Chen, W. High Near-Infrared Reflectance Orange Pigments of Fe-Doped La2W2O9: Preparation, Characterization, and Energy Consumption Simulation. *ACS Sustain. Chem. Eng.* **2021**, *9*, 12385–12393. [CrossRef]

78. Gambelli, A.M.; Cardinali, M.; Filipponi, M.; Castellani, B.; Nicolini, A.; Rossi, F. A normalization procedure to compare retro-reflective and traditional diffusive materials in terms of UHI mitigation potential. *AIP Conf. Proc.* **2019**, *2191*, 020085. [CrossRef]

79. Sakai, H.; Emura, K.; Igawa, N. Reduction of reflected heat by retroreflective materials. *J. Struct. Constr. Eng. (Trans. AIJ)* **2008**, *73*, 1239–1244. [CrossRef]

80. Levinson, R.; Chen, S.; Slack, J.; Goudey, H.; Harima, T.; Berdahl, P. Design, characterization, and fabrication of solar-retroreflective cool-wall materials. *Sol. Energy Mater. Sol. Cells* **2019**, *206*, 110117. [CrossRef]

81. Morini, E.; Castellani, B.; Anderini, E.; Presciutti, A.; Nicolini, A.; Rossi, F. Optimized retro-reflective tiles for exterior building element. *Sustain. Cities Soc.* **2018**, *37*, 146–153. [CrossRef]

82. Rossi, F.; Castellani, B.; Presciutti, A.; Morini, E.; Filipponi, M.; Nicolini, A.; Santamouris, M. Retroreflective façades for urban heat island mitigation: Experimental investigation and energy evaluations. *Appl. Energy* **2015**, *145*, 8–20. [CrossRef]

83. Castellani, B.; Gambelli, A.M.; Nicolini, A.; Rossi, F. Optic-energy and visual comfort analysis of retro-reflective building plasters. *Build. Environ.* **2020**, *174*, 106781. [CrossRef]

84. Manni, M.; Nicolini, A. Optimized Cool Coatings as a Strategy to Improve Urban Equivalent Albedo at Various Latitudes. *Atmosphere* **2021**, *12*, 1335. [CrossRef]

85. Anupam, B.; Sahoo, U.C.; Chandrappa, A.K.; Rath, P. Emerging technologies in cool pavements: A review. *Constr. Build. Mater.* **2021**, *299*, 123892. [CrossRef]

86. Yoon, S.G.; Yang, Y.K.; Kim, T.W.; Chung, M.H.; Park, J.C. Thermal Performance Test of a Phase-Change-Material Cool Roof System by a Scaled Model. *Adv. Civ. Eng.* **2018**, *2018*, 1–11. [CrossRef]

87. Chang, S.J.; Wi, S.; Cho, H.M.; Jeong, S.-G.; Kim, S. Numerical analysis of phase change materials/wood–plastic composite roof module system for improving thermal performance. *J. Ind. Eng. Chem.* **2019**, *82*, 413–423. [CrossRef]

88. Piselli, C.; Castaldo, V.L.; Pisello, A.L. How to enhance thermal energy storage effect of PCM in roofs with varying solar reflectance: Experimental and numerical assessment of a new roof system for passive cooling in different climate conditions. *Sol. Energy* **2019**, *192*, 106–119. [CrossRef]

89. Xie, N.; Li, H.; Zhang, H.; Zhang, X.; Jia, M. Effects of accelerated weathering on the optical characteristics of reflective coatings for cool pavement. *Sol. Energy Mater. Sol. Cells* **2020**, *215*, 110698. [CrossRef]

90. Anupam, B.; Sahoo, U.C.; Rath, P. Phase change materials for pavement applications: A review. *Constr. Build. Mater.* **2020**, *247*, 118553. [CrossRef]

91. Chandel, S.; Agarwal, T. Review of current state of research on energy storage, toxicity, health hazards and commercialization of phase changing materials. *Renew. Sustain. Energy Rev.* **2017**, *67*, 581–596. [CrossRef]

92. Naikwadi, A.T.; Samui, A.B.; Mahanwar, P.A. Fabrication and experimental investigation of microencapsulated eutectic phase change material-integrated polyurethane sandwich tin panel composite for thermal energy storage in buildings. *Int. J. Energy Res.* **2021**, *45*, 20783–20794. [CrossRef]

93. Gong, X.; Wang, C.; Zhu, Q. Research progress on preparation and application of microcapsule phase change materials. *Chem. Ind. Eng. Prog.* **2021**, *40*, 5554–5576.

94. Ling, Z.; Zhang, Y.; Fang, X.; Zhang, Z. Structure effect of the envelope coupled with heat reflective coating and phase change material in lowering indoor temperature. *J. Energy Storage* **2021**, *41*, 102963. [CrossRef]

95. Ferrari, C.; Libbra, A.; Muscio, A.; Siligardi, C. Design of ceramic tiles with high solar reflectance through the development of a functional engobe. *Ceram. Int.* **2013**, *39*, 9583–9590. [CrossRef]

96. Ferrari, C.; Muscio, A.; Siligardi, C.; Manfredini, T. Design of a cool color glaze for solar reflective tile application. *Ceram. Int.* **2015**, *41*, 11106–11116. [CrossRef]

97. Li, Z.; Zhao, M.; Zeng, J.; Wu, J.; Peng, C. High-solar-reflectance building ceramic tiles based on titanite (CaTiSiO$_5$) glaze. *Sol. Energy* **2017**, *153*, 623–627. [CrossRef]

98. Levinson, R.; Akbari, H.; Reilly, J.C. Cooler tile-roofed buildings with near-infrared-reflective non-white coatings. *Build. Environ.* **2007**, *42*, 2591–2605. [CrossRef]

99. Rosado, P.J.; Faulkner, D.; Sullivan, D.P.; Levinson, R. Measured temperature reductions and energy savings from a cool tile roof on a central California home. *Energy Build.* **2014**, *80*, 57–71. [CrossRef]

100. Marangoni, M.; Nait-Ali, B.; Smith, D.; Binhussain, M.; Colombo, P.; Bernardo, E. White sintered glass-ceramic tiles with improved thermal insulation properties for building applications. *J. Eur. Ceram. Soc.* **2016**, *37*, 1117–1125. [CrossRef]

101. Ferrari, C.; Libbra, A.; Cernuschi, F.M.; De Maria, L.; Marchionna, S.; Barozzi, M.; Siligardi, C.; Muscio, A. A composite cool colored tile for sloped roofs with high 'equivalent' solar reflectance. *Energy Build.* **2016**, *114*, 221–226. [CrossRef]

102. Di Giuseppe, E.; Sabbatini, S.; Cozzolino, N.; Stipa, P.; D'Orazio, M. Optical properties of traditional clay tiles for ventilated roofs and implication on roof thermal performance. *J. Build. Phys.* **2018**, *42*, 484–505. [CrossRef]

103. Takebayashi, H.; Moriyama, M.; Sugihara, T. Study on the cool roof effect of Japanese traditional tiled roof: Numerical analysis of solar reflectance of unevenness tiled surface and heat budget of typical tiled roof system. *Energy Build.* **2011**, *55*, 77–84. [CrossRef]

104. Pisello, A.L.; Cotana, F.; Nicolini, A.; Brinchi, L. Development of Clay Tile Coatings for Steep-Sloped Cool Roofs. *Energies* **2013**, *6*, 3637–3653. [CrossRef]

105. Pisello, A.L. Thermal-energy analysis of roof cool clay tiles for application in historic buildings and cities. *Sustain. Cities Soc.* **2015**, *19*, 271–280. [CrossRef]

106. Al Yacouby, A.M.; Khamidi, M.F.; Nuruddin, M.F.; Farhan, S.A.; Razali, A.E. Study on roof tile's colors in Malaysia for development of new anti-warming roof tiles with higher Solar Reflectance Index (SRI). In Proceedings of the National Postgraduate Conference—Energy and Sustainability: Exploring the Innovative Minds, NPC, Perak, Malaysia, 19–20 September 2011; pp. 1–6. [CrossRef]
107. Enríquez, E.; Fuertes, V.; Cabrera, M.; Seores, J.; Muñoz, D.; Fernandez, J. New strategy to mitigate urban heat island effect: Energy saving by combining high albedo and low thermal diffusivity in glass ceramic materials. *Sol. Energy* **2017**, *149*, 114–124. [CrossRef]
108. Cedillo-González, E.; Governatori, M.; Ferrari, C.; Siligardi, C. Solar reflective ink-jet printed porcelain stoneware tiles as an alternative for Urban Heat Island mitigation. *J. Eur. Ceram. Soc.* **2021**, *42*, 707–715. [CrossRef]
109. Rahayu, M.; Sujito; Wibowo, E.; Sutisna, S. Study on the self-cleaning and thermal reducing abilities of TiO_2 coated clay roof tile. *AIP Conf. Proc.* **2021**, *2320*, 030003. [CrossRef]
110. Khaled, K.; Berardi, U. Current and future coating technologies for architectural glazing applications. *Energy Build.* **2021**, *244*, 111022. [CrossRef]
111. Taallah, H.; Chorfa, A.; Tamayo, A.; Rubio, F.; Rubio, J. Investigating the effect of WO_3 on the crystallization behavior of SiO_2–B_2O_3 – Al_2O_3–Na_2O – CaO–ZnO high VIS-NIR reflecting glazes. *Ceram. Int.* **2021**, *47*, 26789–26799. [CrossRef]
112. Kazmi, D.; Williams, D.; Serati, M. Waste glass in civil engineering applications—A review. *Int. J. Appl. Ceram. Technol.* **2019**, *17*, 529–554. [CrossRef]
113. Demir, I. Reuse of waste glass in building brick production. *Waste Manag. Res. J. Sustain. Circ. Econ.* **2009**, *27*, 572–577. [CrossRef]
114. Phonphuak, N.; Kanyakam, S.; Chindaprasirt, P. Utilization of waste glass to enhance physical–mechanical properties of fired clay brick. *J. Clean. Prod.* **2016**, *112*, 3057–3062. [CrossRef]
115. Abdeen, H.H.; Shihada, S.M. Properties of Fired Clay Bricks Mixed with Waste Glass. *J. Sci. Res. Rep.* **2017**, *13*, 1–9. [CrossRef]
116. Silva, R.V.; De Brito, J.; Lye, C.Q.; Dhir, R.K. The role of glass waste in the production of ceramic-based products and other applications: A review. *J. Clean. Prod.* **2017**, *167*, 346–364. [CrossRef]
117. Hameed, A.; Abbas, S.; Qazi, A.; Haider, U. Effect of waste glass on properties of burnt clay bricks. *Pak. J. Eng. Appl. Sci.* **2018**, *22*, 1351.
118. Kazmi, S.M.S.; Munir, M.J.; Wu, Y.-F.; Hanif, A.; Patnaikuni, I. Thermal performance evaluation of eco-friendly bricks incorporating waste glass sludge. *J. Clean. Prod.* **2018**, *172*, 1867–1880. [CrossRef]
119. Gualtieri, M.L.; Mugoni, C.; Guandalini, S.; Cattini, A.; Mazzini, D.; Alboni, C.; Siligardi, C. Glass recycling in the production of low-temperature stoneware tiles. *J. Clean. Prod.* **2018**, *197*, 1531–1539. [CrossRef]
120. Njindam, O.; Njoya, D.; Mache, J.; Mouafon, M.; Messan, A.; Njopwouo, D. Effect of glass powder on the technological properties and microstructure of clay mixture for porcelain stoneware tiles manufacture. *Constr. Build. Mater.* **2018**, *170*, 512–519. [CrossRef]
121. Yinfei, D.; Mingxin, D.; Haibin, D.; Deyi, D.; Peifeng, C.; Cong, M. Incorporating hollow glass microsphere to cool asphalt pavement: Preliminary evaluation of asphalt mastic. *Constr. Build. Mater.* **2020**, *244*, 118380. [CrossRef]
122. Nie, X.; Yoo, Y.; Hewakuruppu, H.; Sullivan, J.; Krishna, A.; Lee, J. Cool White Polymer Coatings based on Glass Bubbles for Buildings. *Sci. Rep.* **2020**, *10*, 6661. [CrossRef]
123. Mourou, C.; Martín-Morales, M.; Zamorano, M.; Ruiz, D.P. Light Reflectance Characterization of Waste Glass Coating for Tiles. *Appl. Sci.* **2022**, *12*, 1537. [CrossRef]

applied sciences

MDPI

Article

Light Reflectance Characterization of Waste Glass Coating for Tiles

Chaimae Mourou [1], María Martín-Morales [1], Montserrat Zamorano [2] and Diego P. Ruiz [3,*]

[1] School of Building Engineering, University of Granada, 18071 Granada, Spain; chaimaemourou@correo.ugr.es (C.M.); mariam@ugr.es (M.M.-M.)
[2] School of Civil Engineering, University of Granada, 18071 Granada, Spain; zamorano@ugr.es
[3] Faculty of Sciences, University of Granada, 18071 Granada, Spain
* Correspondence: druiz@ugr.es

Abstract: Glass wastes that come from recycling plants do not often find a proper use, thus, they are discarded. In order to find future uses for these wastes, this paper explores the characterization of waste glasses (WGs) as a raw material through the assessment of their light reflectance if they were used for external coatings in building materials. To this aim, in this research, several clay-tile specimens were fabricated and coated with three different compositions of waste glass. For these specimens, three variables were analyzed to serve for this WG-based coating characterization: thickness of WG coating, temperature, and holding time of burning. The resulting WG-coated tiles were assessed in terms of the light spectral reflectance and whiteness index, with the help of a fiber optic spectrometer. Results show that the composition of WG had a very significant influence on the light spectral reflectance and the degree of whiteness, with holding time and WG thickness being the most influential depending on the WG type. The temperature of burning was also shown to be critical for the densification process. Finally, an interpretation of these results based on the WG chemical composition coatings obtained by XRF is discussed in this paper.

Keywords: waste glass; light reflectance; building material coatings; soda–lime–silica glass; lead–silica glass

Citation: Mourou, C.; Martín-Morales, M.; Zamorano, M.; Ruiz, D.P. Light Reflectance Characterization of Waste Glass Coating for Tiles. *Appl. Sci.* **2022**, *12*, 1537. https://doi.org/10.3390/app12031537

Academic Editor: Asterios Bakolas

Received: 31 December 2021
Accepted: 29 January 2022
Published: 31 January 2022

Publisher's Note: MDPI stays neutral with regard to jurisdictional claims in published maps and institutional affiliations.

1. Introduction

The remarkable increase of waste glass (WG) generation suggests the study and proposal of more sustainable techniques for waste management and recycling possibilities, which ultimately would benefit the ceramic industry. The adaptation of efficient WG management through the industrial production chain implies environmental gains related to landfilling avoidance, recovery of co-products, and an eco-friendly use of energy through the production process [1]. In accordance with the circular economy principles, an equilibrium should be settled between citizens, municipalities, and solid-waste recycling companies to create a closed-loop supply chain for the co-benefit of all stakeholders [2].

The non-biodegradable nature of glass makes it non-environmental friendly waste [3]; hence, creating new options for recycling WG will alleviate the pressure from both disposal procedures and raw-material extraction. According to the academic literature, promising results have been obtained from recycling WG in the production of eco-friendly ceramic materials regarding the physical, mechanical, and thermal properties [4–9]. However, the light-reflectance evaluation of coating materials containing WG was poorly discussed, especially in the visible and near-infrared radiation (NIR) spectrum [10,11], and this knowledge is relevant for indoor and outdoor environments in terms of human welfare and environmental comfort.

In general, WG was either integrated in substrate materials, such as bricks and tiles, or coating materials, such as glazes, engobes, and binders. For the production of new designs of glazes, the substitution procedure was totally applied in a combined kaolin and bottle

WG for based materials of a new glaze [12], and partially, in the frits composition with a feasible amount of 8% mass as in the case of laminated WG used as a raw material [13]. Moreover, important gains were achieved regarding the cost of the new glazes with WG in their composition [14].

Among WG origins, soda–lime glass, also called soda–lime–silica glass (denoted as SLS), is the most abundant source of glass [15], which explains the growing interest in the research field for their potential use for radiation-shielding applications. The addition of several oxides, such as antimony (III) oxide (Sb_2O_3), improves the radiation-shielding ability of soda–lime–silicate (SLS) WG network [16], and it decreases the X-ray transmission [17]. Moreover, the addition of both lanthanum oxide (La_2O_3) and gadolinium oxide (Gd_2O_3) has the ability to increase the linear attenuation coefficient (LAC) values and, hence, improve its gamma-rays-shielding characteristics [18]. Similar results were obtained with the addition of MoO_3 [19]. Moreover, it is frequently found in WGs for their shielding applications and their optical properties the flint glass or lead silicate glass (denoted as LS), which contains a minimum of 24% (by weight) lead (II) oxide (PbO). A broad range of scientific studies used chelating treatments to recycle LS WG resulting from cathode ray tubes (CRT) [20,21], since the amounts of CRT wastes have increased around the world after gradually replacing it with liquid crystal displays (LCD) [22]. CRT glass could be considered as a substitute for non-plastic materials, in particular, ceramic frits obtained from mixtures of silicates and carbonates to produce ceramic glazes [23]. On the other hand, and in contrast to SLS WG, the lead silicate (LS) WG requires much more caution both in recycling and in the disposal measures, since it contains lead metal.

The objective of this work is to establish a first proof of concept related to the identification of the influence in both the specular light spectral reflectance and the degree of whiteness of the WG coated tile specimens, taking into account the type of WG materials and the manufacturing process characterized by the parameters of temperature of burning, thickness of WG coating, and time of burning.

In this regard, it is interesting to assess the light-reflectance properties of the WG coated tiles if they are planned to be used as coatings on tiles located in roofs, walls, or other applications in the construction sector. The characterization of solar-reflectance performance of WG coatings, especially in the near-infrared, was poorly discussed in the literature, and even less evidence on the use of waste glass as coatings on tiles. Therefore, a first proof of concept of this influence was performed by testing the relevance of variables, such as the WG-coated tile's composition, and other manufacturing features, such as the holding time or temperature of burning in the whiteness index or light spectral reflectance.

To achieve this goal, laboratory measurements were used to determine the light spectral reflectance and degree of whiteness of the specimens in the wavelength range of 350–1100 nm. The overall process was divided into four main steps: characterization of raw materials, production of the specimens, firing stage, and, finally, several tests for the measurement of the degree of whiteness and the specular light reflectance were conducted. Therefore, this work aimed at the use of three different types of WG in the preparation of coating for clay tiles specimens: two types of SLS WGs with different compositions and an LS WG derived from CRT.

2. Materials and Methods

This research aimed to study the behavior of glazed clay surfaces made of WG in terms of their light-spectral-reflectance performance. In the experimental setup, several WG-coated tile specimens were manufactured by using three values of WG thickness (0.5, 0.75, and 1 mm), with different burning temperatures (700, 850, and 1000 °C), using the laboratory kiln with different holding times (20, 40, and 60 min). In this section, the materials and their manufacturing based on the temperature of burning, thickness of WG coating, and time of burning are described. It is also described the experimental setup used to measure the specular light spectral reflectance and the degree of whiteness of the tile samples.

2.1. Origin of Materials

In this work, three types of WG were used for coatings. The WGs were provided by the company "Camacho Recycling", and they were taken from the glass-collection plant located in Albacete, Spain. This company has developed glass-collection systems for all the different types of glass, regardless of the origin, composition, and quantities that can be generated on factories or homes. The preparation of the substrate (ceramic body) was carried out by using a clay powder provided by the local company "Ladrillos Suspiro del Moro S.L" in Granada, Spain, under the instructions and supervision of the research team. Both companies have a long history in the collection and recycling of waste glass, and the manufacturing of bricks or other construction materials.

2.2. Preparation of the Flat-Tile Specimens

According to the dosage provided by the company, 81 clay substrates of 3.2 mm × 3.2 mm × 1.5 mm were fabricated by mixing clay powder with 15 wt.% of water ("wt." stands for percentage of weight per unit volume). The production process is shown in Figure 1, and it can be summarized as follows:

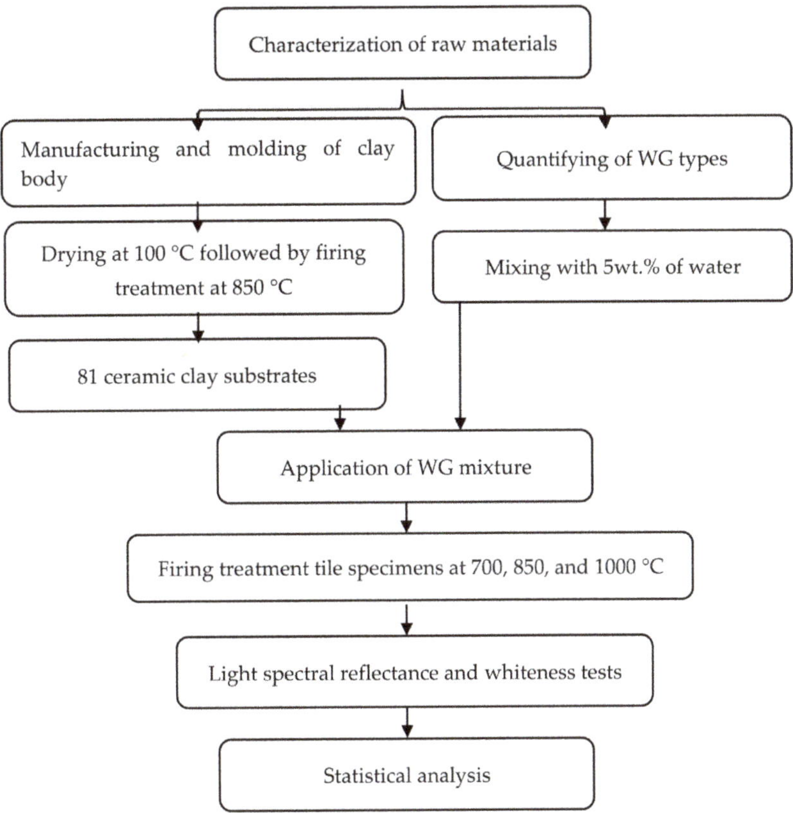

Figure 1. Flowchart of the research procedure.

Firstly, clay powder was processed and treated by a drying process until its weight was stabilized. Then the process was performed on a homogenous mixture of the resulted clay powder and water with the help of an electrical mixer. The molding of the substrates was achieved with the help of a mold (Figure 2) fabricated according to the dimensions required for the further tests' measurements. With the help of a compressive test machine, uniaxial pressing was applied on the mold to shape the clay body at 1000 MPa. This process was followed by a drying for 48 h under the temperature of 100 °C at the laboratory furnace, and then a firing treatment during 1 h under the temperature of 850 °C at the laboratory kiln. In this work, we opted for a double firing process; the biscuit state was obtained by drying the 81 clay substrates for 48 h at 100 °C, using the laboratory furnace, followed by a firing treatment during 1 h at 850 °C, using the laboratory kiln.

Figure 2. Mold used in the laboratory for the specimen shaping.

In a second step, the coating was added to the clay substrate. According to the varying parameters of the study, for the preparation of WG coatings, three quantities of each WG type representing the three values of thickness were mixed with 5 wt.% of water (Table 1) in order to obtain a mixture that could be spread evenly enough to achieve the desired thicknesses. The next step was the manual addition of the mixture on the fired-clay body. Finally, the heat treatment was processed according to two variables: holding time and temperature, i.e., 20, 40, and 60 min for each of the temperatures of 700, 850, and 1000 °C.

Table 1. Identification of WG specimens.

Waste Glass (WG) Type	Particle Size (mm)	Identification	No. of Specimens
1st SLS WG	(0, 1)	WG1_Qx [1]_y [2]	27
2nd SLS WG	(0–3)	WG2_Qx_y	27
LS WG	(0–4)	WG3_Qx_y	27

[1] Thickness of WG coating (Q1 = 0.5, Q2 = 0.75 and Q3 = 1 mm). [2] Time of burning: 20, 40, and 60 min.

In summary, several samples of every coated tile were manufactured, and three samples for each one of the cases with different temperatures, holding time, and thickness were selected to be studied and used for the light spectral reflectance and whiteness measurements. In total, 27 samples of each WG type were chosen (Table 1).

2.3. Chemical Characterization of Raw Materials and Clay Specimens Using XRF

The characterization of the three WG types and the clay powder was conducted through the chemical composition obtained by X-ray fluorescence (XRF) analysis. This non-destructive test method is used to analyze the structure of the clay samples and reveal their chemical composition. In the XRF-based analysis, a primary X-ray beam was directed at a sample, and we measured the secondary X-ray emitted from a sample (called fluorescence) when it is excited by the primary X-ray source. Every element in a sample produces a set of unique characteristic fluorescent X-rays that allows us to determine the chemical composition of materials. The equipment used was a Philips MagiX 2400. The equipment was calibrated with the corresponding standard sample. The analysis of the majority elements was carried out by preparing a bead by mixing 0.3 g of sample and 5.5 g of Lithium Tetraborate. Quantification was carried out by using the quantitative analysis curve for silico-aluminous materials. When the concentration of the elements was low, i.e., they were present in trace form, the pressed tablet or pellet method was used.

2.4. Surface Spectral Reflectance Measurement of WG Coated Tile Samples

In this research, a spectrometer was used to perform the measurement of the specular light spectral reflectance of surfaces. For this type of reflectance measurements, the StellarNet miniature spectrometers family members are suitable, since they are a portable and compact fiber optic instruments for ultraviolet, visible (VIS), and near infrared (NIR) measurements offering CCD 2048 and PDA 512/1024 detectors with the required accuracy for the objectives of this research.

Specifically, the experimental setup was as follows (see Figure 3). For the measurements, a StellarNet BLUE-Wave Spectrometer of STN–BW–VIS type was used. This spectrometer is a fiber-optic-coupled instrument for measurements in the range of 350–1150 nm wavelength. It uses a 16-bit digitizer via high speed USB-2, and each unit contains a USB-2 interface with a snap shot memory to provide instantaneous spectral image from the highly sensitive CCD or Photo Diode Array detectors. The reflectance probe used was a STN-R600-8-VisNIR type for VIS and NIR (400–2200 nm wavelength) measurements. This probe was assembled in a reflectance probe holder for 90° angle measurements and this strand fiber optic cable or probe assembly delivers input via standard SMA 905 fiber optic connector. The experimental setup also contains a light source STN-SL1 type, which is a 10,000 h Tungsten and Halogen lamp, 2800 Kelvin color temperature, 350–2500 nm (Figure 3). This spectrometer equipment was calibrated with NIST (National Institute of Standards and Technology) traceability.

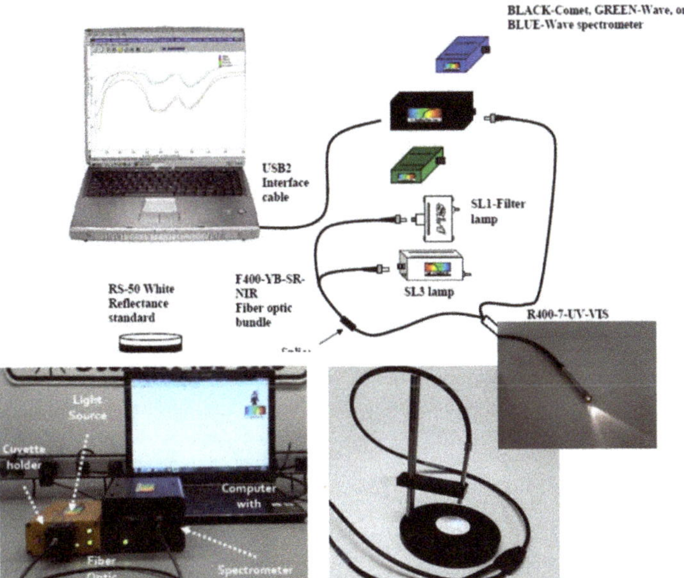

Figure 3. StellarNet BLUE-Wave Spectrometer STN–BW–VIS with the reflectance probe and probe holder for the light spectral reflectance measurements.

The equipment also contains the STN-RS50 reflectance standard, which is a 50 mm diameter white reflectance standard made of Halon. It is used to take reference measurements by using the R600-8 Reflectance Probe. The white standard will reflect >97% of the light from 300 to 1700 nm. Data were recorded by using the SpectraWiz software to accurately measure the light reflected intensity and perform other spectral calculations.

Once the WG-coated tile specimen was placed in the sample holder in a dark lab room, the experimental reflectance data-collection procedure became as follows:

- Dark spectrum measurement: it records the background noise with the source turned off. Dark spectrum is subtracted from measurements.
- Reference spectrum measurement: it records the reference spectrum with the STN-RS50 white reflectance standard.
- Sample spectrum measurement: it records the quotient between the sample reflectance spectrum and the reference spectrum of the RS50 standard.

For each reflectance measurement of the WG coated samples, the number of spectra to signal averaging was set. This option provides a smoothing effect, thus increasing the system signal-to-noise ratio by the square root of the number of scans being averaged. The rule is to set the averaging to the highest number tolerable when there is sufficient light signal, keeping the detector integration time short but out of saturation. In our measurements, the integration time was kept above 30 ms and at least 10 scans were averaged.

Since we manufactured three WG-coated tile samples with the same characteristics of holding time, temperature, and thickness, as stated in Section 2.2, the experimental data collection was performed for the 27 samples, and each measurement was repeated three times with each sample to ensure the quality of acquired data.

For all the samples, we measured the specular light spectral reflectance in the visible spectral range that extends from 400 to 700 nm and the NIR spectral range from 700 to 1100 nm. We also computed the degree of whiteness or whiteness index denoted as L* according the CIELAB D65 reference of the French-based international Commission on Illumination (CIE). This CIE whiteness index is a single number, which references the relative degree of near white materials under specific lighting conditions, and it correlates the visual ratings of whiteness for certain surfaces compared to the white-surface standard in the visible spectrum range. L* increases with whiteness, reaching for our applications the maximum value of 100 for the perfect white sample [24].

Finally, with the aim of gaining some insight of data, some measurements were processed through ANOVA analysis, using SPSS software. The statistical analysis focused on analyzing the existence of differences between the probability distributions for the different spectral reflectance measurements or whiteness index measurements of the WG coated tile samples. The normality of the data was checked by goodness-of-fit tests (P–P probability plots or the Kolmogorov–Smirnov test). In our cases, the data distributions were not normal, so the non-parametric Mann–Whitney U test was used in all cases. The results were interpreted following the specific statistical analysis.

3. Results

In this section, we provide the results obtained for the WG-coated tile samples based on the effect of variables such as temperature of burning, thickness of WG coat, and time of burning in the specular light spectral reflectance and whiteness index, following the methodology described in the preceding section. We also give the experimental chemical characterization of materials of the tile samples.

3.1. Chemical Characterization of Raw Materials

WG particles used in this study were classified into three types, based on their origins, grain sizes, and chemical compositions (Table 2). The XRF chemical characterization of the glass shows that it contains SiO_2; fluxing elements, such as Na_2O, K_2O, and PbO; and stabilizing elements, such as Al_2O_3, CaO, BaO, and MgO. The first type, denoted as WG1, is a hollow green glass that was collected from recycled bottles, and, according to its composition, it is an SLS type of glass. The second type (denoted as WG2) contains some flat glass, and it is an SLS-type glass, with the addition of a small quantity of stones and ceramic materials. The third type (denoted as WG3) is an LS glass mainly coming from CRT TV monitors. The clay powder was obtained by mixing two types of raw constituents, namely 40% grea and 60% lime. To eliminate the moisture, the samples were oven-dried at 100 °C for 24 h to constant mass.

Table 2. Average chemical composition in wt.% of WG and clay obtained from FRX analysis. Please note that they are average values and do not necessarily add up to 100% for each element.

	SiO_2	Na_2O	CaO	MgO	Al_2O_3	Fe_2O_3	K_2O	TiO_2	P_2O_5	BaO	PbO
Clay	44.46	0.69	11.21	3.63	16.08	5.48	3.34	0.66	0.14	-	-
WG1 [1]	73.2	11.4	10.8	1.35	2.03	0.31	0.88	0.066	<0.04	-	-
WG2 [2]	72	13	9	2	1.75	<0.1	0.55	-	-	-	
WG3 [3]	52.5	6	3	1.75	2.25	0.15	7.5	0.075	-	2	20.5

[1] SLS WG with a particle size between (0 and 1) mm. [2] SLS WG with a particle size between (0 and 3) mm. [3] LS WG with a particle size between (0 and 4) mm.

3.2. Qualitative Visual Characterization of the WG Coated Specimens

As a result of the firing treatment, none of the WG types reached the melting point up to the temperature of 700 °C. At 850 °C, WG1 and WG2 types created a dense coat that did not adhere to the substrate, whereas WG3 showed more densification; however, it did not cover the entire surface. At 1000 °C, all types of WG reached the melting point, and a dense coat structure was formed. The tiles specimens fired at 1000 °C showed better properties in terms of material adhesion when compared to the 700 and 850 °C temperature cases. Hence, this temperature was set up, and according to the structure and aesthetical perspective, 27 specimens fired at 1000 °C (Figure 4a–c) were selected for further tests. All the samples had a homogeneous visual appearance to ensure the reproducibility of data.

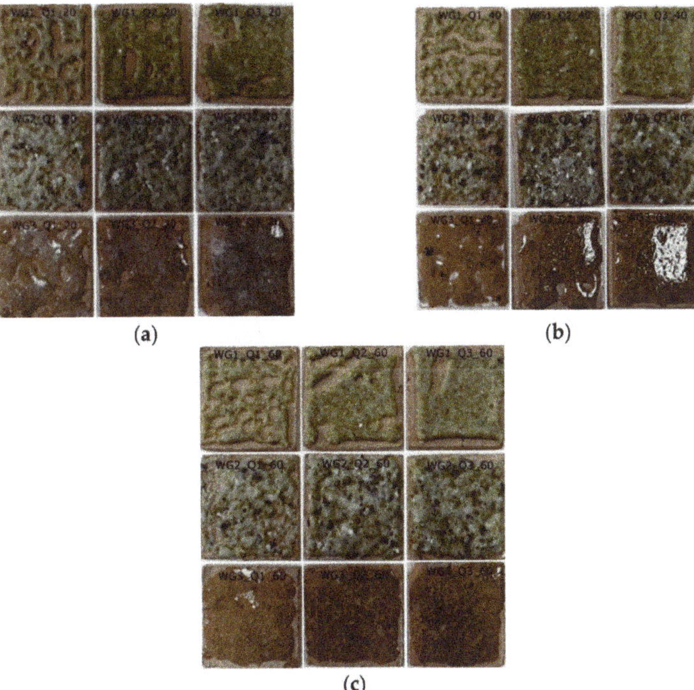

Figure 4. Tiles specimens fired at 1000 °C during (**a**) 20 min, (**b**) 40 min, and (**c**) 60 min. In each figure, the first row shows WG1, the second row WG2, and the third row WG3 coated samples respectively. In addition, in each figure, the first column shows Q1 samples, the second column shows Q2 samples, and the third column shows Q3 samples according to Table 1.

The coating structure of the three types of WG was different. According to Figure 4a, WG1 presented a weak transition zone characterized by some cracking, and an unevenly distribution of melted WG due to particle shrinkage for small quantity of WG that improved with the increase of WG quantity. WG2 showed a porous surface and more roughness than the other types, and these qualities were notably reduced at a holding time of 60 min (Figure 4c).

WG1 and WG2 coatings had the appearance of an opaque surface lying on the substrate, but the WG3-coated samples had a glossy surface that is notably transparent; in fact, some breaks or rifts can be observed at the transition zone between the substrate and the coating. Moreover, WG3 coating contained air bubbles, and some pinholes occurred due to reactions of the oxides (Figure 4a). These effects decreased with the increase of the holding time (Figure 4c).

3.3. Light Characterization of the WG-Coated Specimens

Measurements of the whiteness index L* and light spectral reflectance measurements were taken for the 27 specimens described above. Table 3 shows the value of L* for the three WG-coated samples. In this table, WG1Q1 stands for the sample with a coating thickness of Q1 (Table 1) and so on. As can be observed in this table, the whiteness index ranged from 50.45 to 52.93, with an average value of 51.78, for WG1; it ranged from 49.24 to 55.23, with an average value of 52.93, for WG2; and it ranged from 54.54 to 100, with an average value of 84.49, for WG3.

Table 3. Degree of whiteness (L*) of the 27 WG-coated tile samples.

Burning Time	L*								
	WG1Q1	WG1Q2	WG1Q3	WG2Q1	WG2Q2	WG2Q3	WG3Q1	WG3Q2	WG3Q3
20 min	51.76	50.69	50.56	55.23	52.56	51.28	61.39	97.33	100
40 min	52.93	52.18	52.16	53.86	54.8	53.25	71.46	83.22	100
60 min	52.41	50.45	52.88	49.24	54.78	51.45	54.54	92.5	100

An ANOVA analysis according to the methodology section was also performed on the data coming from Table 3, and all the results can be found in Appendix A. The statistical results coming from the nonparametric Mann–Whitney U test showed the following:

- In general, the variables "holding time of burning" and "thickness of WG coating" had no significant influence on the whiteness index.
- For the type of WG, the results showed no significant difference between the WG1 and WG2 types, in contrast to the WG3, which had a significant influence. In other words, the results show that there was no significant difference between the mean values of the whiteness index for the WG1- and WG2-coated types, and only the degree of whiteness of the WG3 type was significantly influenced at a $p < 0.05$ level.
- The WG3 samples registered the highest values of the whiteness index.

Regarding the relative light spectral reflectance measurements, Figures 5–7 show the different light spectra reflectance obtained for the three WG-coated types of samples.

In this case, the influence of the holding time of burning and the WG-coated types on the spectral reflectance measurements was checked, since, from the manufacturing point of view, it is very interesting to know if there is an influence of the holding time on the spectral reflectance properties. Each figure represents the relative specular spectral reflectance of the WG1-, WG2-, and WG3-coated samples (according to Table 2) with different holding burning times. For example, WG1T20 stands for the WG1-coated sample measurements with a holding time of burning of 20 min, and WG2T40 stands for the WG2-coated sample with a holding time of burning of 40 min, and so on. Figure 5 shows the spectral reflectance data obtained for the 20 min burning time, Figure 6 for the 40 min burning time, and Figure 7 for the 60 min burning time.

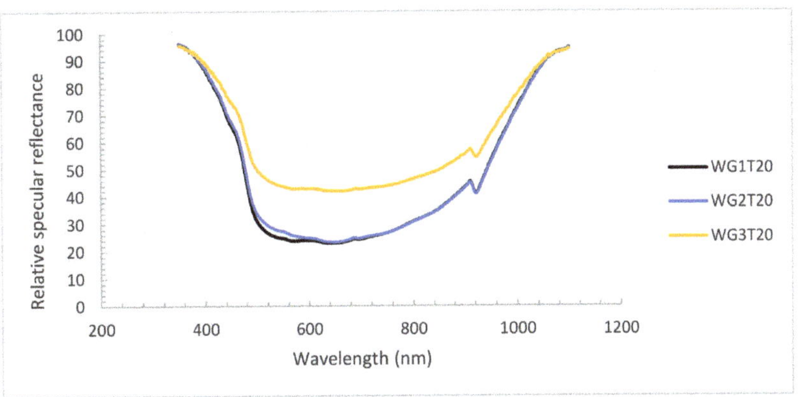

Figure 5. Specular light reflectance of the first set of specimens fired at 1000 °C during 20 min.

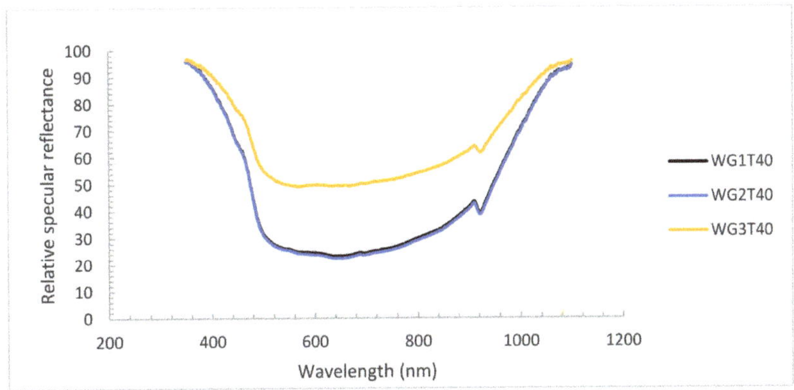

Figure 6. Specular light reflectance of the first set of specimens fired at 1000 °C during 40 min.

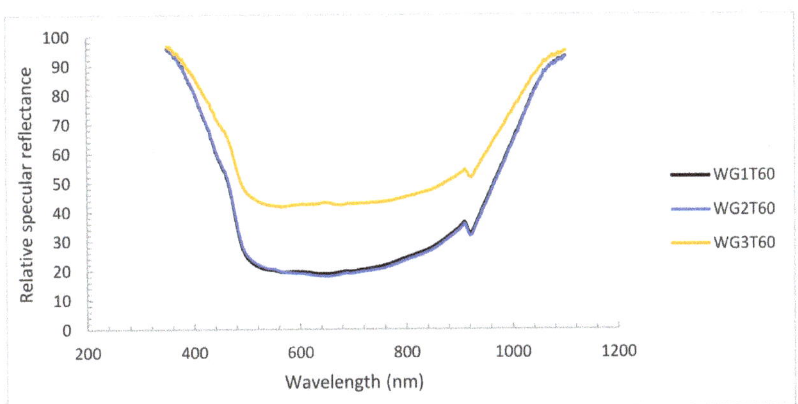

Figure 7. Specular light reflectance of the first set of specimens fired at 1000 °C during 60 min.

As it can be observed from these figures, all types of WG-coated tiles reached a minimum light reflectance in the range of 500–800 nm, which tends to increase through the rest of the visual range and then increases gradually in the NIR range. The region in the range of 500–800 nm, which roughly corresponds to the visible range (yellow, orange, and red colors), is the one showing almost constant and minimum values of the relative light reflectance. Therefore, these values are the most interesting ones to look for differences between the behavior of the samples on the global light reflectance, and they were chosen to perform a series of ANOVA tests, whose results are shown in Appendix B. In this case, the aim was to determine if the spectral data are really influenced by the WG type of glass coating or the holding times of burning.

From the nonparametric Mann–Whitney U tests for the specular light spectral reflectance in terms of the WG type of glass coating and time of burning shown in Appendix B, the following can be concluded:

- There was a significant difference between all the types of WG coatings (WG1, WG2, and WG3) regardless of the holding time ($p < 0.001$).
- There was a significant difference between the holding burning times "20 and 60 min" and "40 and 60 min" of WG1 with $p < 0.001$.
- There was no significant difference between the holding times "20 and 40 min" of WG1 with $p > 0.01$.
- There was a significant difference between all the holding times of WG2 and WG3 with $p < 0.001$.
- The WG3 type had the higher mean values, showing a greater reflection capacity compared to the other types.

In summary, this proof of concept about the spectral reflectance values of the WG coating showed that the light spectral reflectance behavior is highly influenced in the spectrum range that extends from 400 to 800 nm for the composition of the three types of glass used in the WG coatings. The holding time of burning also had some influence, but it was much smaller in general, except in the WG1 type between the 20 and 40 min burning times, wherein no difference was reported.

4. Discussion

The WG coatings tested in this research showed different behaviors during the sintering process. In this regard, the chemical and qualitative visual characterizations of the WG-coated specimens showed clear distinguishing features, and, consequently, it was expected that they would exhibit different values for the whiteness index and the light spectral reflectance. This consideration was checked by a set of experiments, and, thus, it was studied if a relevant different behavior for the WG-coated tiles in terms of both manufacturing feasibility and reflected radiant energy properties would be observed, taking into account their different WG composition and thickness of the coating, as well as other external manufacturing features, such as the holding times or temperatures of burning.

Focusing on the manufacturing process it was interesting to study two variables: holding time and temperature of burning. In our research, we used the holding times of 20, 40, and 60 min for each of the burning temperatures of 700, 850, and 1000 °C. The three different types of WG used in this study showed different visual and physical structures after the firing treatment, with a better densification occurring at 1000 °C with respect to the other tested temperatures. Hence, this temperature was finally selected to build the WG-coated tiles for the next steps of the experiments.

Once this temperature was chosen and the WG coated tiles were manufactured, the degree of whiteness or whiteness index was measured using an experimental setup based on light spectrometry. The composition of WG (that defines the type) in this study, had a significant influence on the whiteness index (L*) of the specimens. WG3 specimens had the highest mean value of L* (84.5) compared to the other types. This can be interpreted as that the small percentages of iron oxide increase the whiteness of the specimens, in accordance to other applications considered in [25] where the whiteness indicator of an engobe containing WG was influenced by the ratio of kaolin, alumina and zirconium. However, in contrast to the WG type or composition, in most cases, the holding time for the firing treatment and the thickness of the WG coating had a small influence on L*, especially for SLS WG specimens.

The third set of experiments considered the two most relevant variables (WG type and holding time) for checking for differences in the specular light spectral reflectance. In terms of light-reflectance measurements, it was measured the relative reflectance in the visible range from 350 to 700 nm and the NIR region from 700 to 1100 nm. In this case, a set of WG1, WG2 and WG3 tiles obtained using three burning times of 20, 40 and 60 min respectively, were considered for testing if there are differences in the spectral reflectance.

From the obtained results, the composition of WG (that defines the type) in this study had again a relevant and significant influence on the light spectral reflectance of the specimens. Holding times had influence mainly in the reflectance properties for the WG2 and WG3 types, generally decreasing the values of reflectance but with a weak influence compared to the strong dependence with the WG type. For all the WG-coated tiles, WG3 had greater values of reflectance with respect to WG1 and WG2. This can be explained from its chemical composition, where the existence of substances with high refractive indexes such as Al_2O_3 and TiO_2 enhances the light reflectance of specimens. In fact, this effect was also reported in other applications where the use of WG in the production of engobes and glazes compositions increased the refractive indexes [10,11]. According to the visual characterization, the glossy transparent structure of WG3 enhanced the light reflectance (Figure 4), based on the LS composition which has a high refractive index, and less internal friction with respect to SL glass [26].

In summary, these results are interpreted based on the chemical composition and the manufacturing process itself. With the increase of waste glass quantity, the degree of whiteness of lead silicate waste glass (LS WG) registered the highest values with respect to soda lime silicate glass (SLS WG). For the relative light reflectance, the lead silicate glass showed better performance with respect to the soda lime silicate glass due to its chemical composition. The difference of the light reflectance performance of the specimens could be explained by the percentage of crystalline phase developed through the sintering process of each type of WG, which makes the degree of whiteness results in agreement with the light reflectance measurements.

5. Conclusions

This research paper presented a preliminary study on using waste glass as a coating on tiles in terms of light reflectance. An optical characterization in terms of light-reflectance properties of the WG-coated tiles is provided by testing the relevance of variables, such as the WG-coated tile composition and other manufacturing features, such as the holding time or temperature of burning in the whiteness index or light spectral reflectance.

From a set of experiments based on raw WG coming from recycling plants, it was shown that the chemical composition and quantity of waste glass, as well as the temperature and holding time, had a significant influence on the light spectral reflectance of the specimens. The same results were found for the degree of whiteness, except from the holding burning time variable that had a minor impact. In general, the LS glass mainly coming from CRT TV monitors (WG3) had the highest mean value of the whiteness index and the higher light reflectance compared to the ones built with an SLS type of glass (WG1 and WG2). The holding time for the firing treatment and the thickness of WG coatings had a smaller influence on the whiteness index, especially weaker for the SLS WG specimens. Holding times also had an influence mainly in the reflectance properties for the LS-glass-coating type, generally decreasing the values of reflectance as the burning time increases.

Some potential benefits can be derived from this study to be considered for the use of SLS and LS waste glass in the production of coatings for clay tile specimens. The production of glazes as glass-ceramics is a result of a controlled crystallization process to fulfill the requirements of a prescribed design. Previous studies investigated the use of WG for the replacement of raw materials used in the production of glazes, and this replacement yielded promising results reducing energy consumption [12–14]. This is confirmed in this research, where a total replacement is carried out through the production of a WG coating material adopting sustainable routes as far as possible, ensuring a consistent seal and compatibility of the substrate and the coating if they were used in the construction sector. Moreover, it was shown that this use of WGs as coating affects the whiteness and the light reflectance performance of tiles, and so it becomes relevant the assessment of their potential impact in the outdoor or indoor environment in buildings, as well as cities. Depending on the specific outdoor environment, one or other coating types would be preferable. For example, if the application was aimed for cool roofs to reflect more sunlight and absorb less solar energy, it would be more suitable to use LS WG-coated tiles with one hour of holding times. Thus, the results obtained could be applied for the cool roof strategy as a main application that will provide, at the greatest extent, waste recovery and energy efficiency in buildings. More potential applications, such as their use in walls and pavements, will be further evaluated in future research.

Author Contributions: C.M. and M.M.-M., conceptualization, performed the experiments, writing the manuscript, formal analysis, and carried out the post-processing; M.Z. and D.P.R., conceptualization, writing the manuscript, project administration, funding acquisition, and supervising the manuscript. All authors have read and agreed to the published version of the manuscript.

Funding: This research was funded by the State Research Agency (AEI) of Spain and European Regional Development Funds (ERDF) under project PID2019-108761RB-I00.

Institutional Review Board Statement: Not applicable.

Informed Consent Statement: Not applicable.

Data Availability Statement: Data are available on request.

Acknowledgments: The authors wish to thank "Camacho Recycling" company for providing the samples used for this research.

Conflicts of Interest: The authors declare no conflict of interest.

Appendix A. Statistical Analysis for the Whiteness Index

Table A1. ANOVA for the whiteness of the specimens.

Factor	Model	Sum of Squares	df	Mean Square	F	Sig.
WG type	Linear	6201.564	2	3100.782	28.708	0.000
holding time	Linear	15.205	2	7.603	0.021	0.979
thickness	Linear	729.768	2	364.884	1.086	0.354

Table A2. Ranks table of Mann–Whitney U test for the whiteness, according to the holding time.

	Ranks			
	Holding Time	**N**	**Mean Rank**	**Sum of Ranks**
	Holding time of 20 min	9	8.39	75.50
	Holding time of 40 min	9	10.61	95.50
	Total	18		
Whiteness	Holding time of 20 min	9	9.83	88.50
	holding time of 60 min	9	9.17	82.50
	Total	18		
	Holding time of 40 min	9	10.61	95.50
	holding time of 60 min	9	8.39	75.50
	Total	18		

Table A3. Test statistics table of Mann–Whitney U test for the whiteness, according to the holding time.

	Test Statistics		
	Whiteness		
	20 min	40 min	60 min
Mann–Whitney U	30.500	37.500	30.500
Wilcoxon W	75.500	82.500	75.500
Z	−0.883	−0.265	−0.883
Asymp. Sig. (2-tailed)	0.377	0.791	0.377
Exact Sig. [2*(1-tailed Sig.)]	0.387	0.796	0.387

Table A4. Ranks table of Mann–Whitney U test for the whiteness according to the thickness.

	Ranks			
	Thickness	N	Mean Rank	Sum of Ranks
	Q1	9	8.89	80.00
	Q2	9	10.11	91.00
	Total	18		
	Q1	9	9.67	87.00
Whiteness	Q3	9	9.33	84.00
	Total	18		
	Q2	9	9.33	84.00
	Q3	9	9.67	87.00
	Total	18		

Table A5. Test statistics table of Mann–Whitney U test for the whiteness, according to the thickness.

	Test Statistics		
	Whiteness		
	Q1	Q2	Q3
Mann–Whitney U	35.000	39.000	39.000
Wilcoxon W	80.000	84.000	84.000
Z	−0.486	−0.133	−0.133
Asymp. Sig. (2-tailed)	0.627	0.894	0.894
Exact Sig. [2*(1-tailed Sig.)]	0.666	0.931	0.931

Table A6. Ranks table of Mann–Whitney U test for the whiteness, according to the WG type.

	Ranks			
	WG	N	Mean Rank	Sum of Ranks
	WG1	9	7.56	68.00
	WG2	9	11.44	103.00
	Total	18		
	WG1	9	5.00	45.00
Whiteness	WG3	9	14.00	126.00
	Total	18		
	WG2	9	5.33	48.00
	WG3	9	13.67	123.00
	Total	18		

Table A7. Test statistics table of Mann–Whitney U test for the whiteness according to the WG type.

	Test Statistics		
	Whiteness		
	WG1	WG2	WG3
Mann–Whitney U	23.000	0.000	3.000
Wilcoxon W	68.000	45.000	48.000
Z	−1.545	−3.584	−3.318
Asymp. Sig. (2-tailed)	0.122	0.000	0.001
Exact Sig. [2*(1-tailed Sig.)]	0.136	0.000	0.000

Appendix B. Statistical Analysis for the Specular Solar Reflectance

Table A8. ANOVA for the specular light reflectance of the specimens.

Factor	Model	Sum of Squares	df	Mean Square	F	Sig.
WG type	Linear	4153.132	2	2076.566	22,159.150	0.000
holding time	Linear	177.840	2	88.920	69.203	0.000

The specular solar reflectance of all types of WG does not follow a normal distribution; hence, the Mann–Whitney U test was carried out. A significant difference was found between all types of WG during each holding time, as shown in Appendix B Tables A9–A14 with a $p < 0.001$.

Table A9. Ranks table of the Mann–Whitney U test for specular solar reflectance, according to the holding time of 20 min.

		Ranks		
	WG	N	Mean Rank	Sum of Ranks
	WG1T20	607	531.00	322.31450
	WG2T20	607	684.00	415.19050
	Total	1214		
	WG1T20	607	304.00	184.52800
	WG3T20	607	911.00	552.97700
	Total	1214		
Reflectance	WG2T20	607	304.00	184.52800
	WG3T20	607	911.00	552.97700
	Total	1214		

Table A10. Test statistics table of the Mann–Whitney U test for specular solar reflectance, according to the holding time of 20 min.

	Test Statistics		
	Reflectance		
	WG1T20_WG2T20	WG1T20_WG3T20	WG2T20_WG3T20
Mann–Whitney U	137.786500	0.000	0.000
Wilcoxon W	322.314500	184.528000	184.528000
Z	−7.605	−30.170	−30.168
Asymp. Sig. (2-tailed)	0.000	0.000	0.000

Table A11. Ranks table of the Mann–Whitney test for specular solar reflectance, according to the holding time of 40 min.

	Ranks			
	WG	**N**	**Mean Rank**	**Sum of Ranks**
	WG1T40	607	669.97	406.67300
	WG2T40	607	545.03	330.83200
	Total	1214		
	WG1T40	607	304.00	184.52800
Reflectance	WG3T40	607	911.00	552.97700
	Total	1214		
	WG2T40	607	304.00	184.52800
	WG3T40	607	911.00	552.97700
	Total	1214		

Table A12. Test statistics table of the Mann–Whitney U test for specular solar reflectance according to the holding time of 40 min.

Test Statistics			
Reflectance			
	WG1T40_WG2T40	**WG1T40_WG3T40**	**WG2T40_WG3T40**
Mann–Whitney U	146.304000	0.000	0.000
Wilcoxon W	330.832000	184.528000	184.528000
Z	−6.210	−30.169	−30.168
Asymp. Sig. (2-tailed)	0.000	0.000	0.000

Table A13. Ranks table of Mann–Whitney U test for specular solar reflectance, according to the holding time of 60 min.

	Ranks			
	WG	**N**	**Mean Rank**	**Sum of Ranks**
	WG1T60	607	663.64	402.83250
	WG2T60	607	551.36	334.67250
	Total	1214		
	WG1T60	607	304.00	184.52800
Reflectance	WG3T60	607	911.00	552.97700
	Total	1214		
	WG2T60	607	304.00	184.52800
	WG3T60	607	911.00	552.97700
	Total	1214		

Table A14. Test statistics table of Mann–Whitney test for specular solar reflectance, according to the holding time of 60 min.

Test Statistics			
Reflectance			
	WG1T60_WG2T60	**WG1T60_WG3T60**	**WG2T60_WG3T60**
Mann–Whitney U	150.144500	0.000	0.000
Wilcoxon W	334.672500	184.528000	184.528000
Z	−5.583	−32.253	−32.248
Asymp. Sig. (2-tailed)	0.000	0.000	0.000

Appendix B Tables A15–A20 show a significant difference between the holding times "20 and 60 min" and "40 and 60 min" of WG1 with $p < 0.001$. While there was no significant difference between the holding times "20 and 40 min" of WG1 with $p = 0.046$.

There was a significant difference between all the holding times of WG2 and WG3 with $p < 0.001$.

Table A15. Ranks table of Mann–Whitney U test for specular solar reflectance of WG1 during the three holding times.

	Ranks			
	WG	**N**	**Mean Rank**	**Sum of Ranks**
	WG1T20	607	587.45	356.58400
	WG1T40	607	627.55	380.92100
	Total	1214		
Reflectance	WG1T20	607	895.54	543.59200
	WG1T60	607	319.46	193.91300
	Total	1214		
	WG1T40	607	898.77	545.55300
	WG1T60	607	316.23	191.95200
	Total	1214		

Table A16. Test statistics table of Mann–Whitney U test for specular solar reflectance of WG1 during the three holding times.

	Test Statistics		
	Reflectance		
	WG1T20_WG1T40	**WG1T20_WG1T60**	**WG1T40_WG1T60**
Mann–Whitney U	172.056000	9385.000	7424.000
Wilcoxon W	356.584000	193.913000	191.952000
Z	−1.993	−28.636	−28.955
Asymp. Sig. (2-tailed)	0.046	0.000	0.000

Table A17. Ranks table of Mann–Whitney U test for specular solar reflectance of WG2 during the three holding times.

	Ranks			
	WG	**N**	**Mean Rank**	**Sum of Ranks**
	WG2T20	607	719.24	436.57750
	WG2T40	607	495.76	300.92750
	Total	1214		
Reflectance	WG2T20	607	902.96	548.09850
	WG2T60	607	312.04	189.40650
	Total	1214		
	WG2T40	607	889.15	539.71450
	WG2T60	607	325.85	197.79050
	Total	1214		

Table A18. Tests statistics table of Mann–Whitney U test for specular solar reflectance of WG2 during the three holding times.

	Test Statistics		
	Reflectance		
	WG2T20_WG2T40	**WG2T20_WG2T60**	**WG2T40_WG2T60**
Mann–Whitney U	116.399500	4878.500	13.262500
Wilcoxon W	300.927500	189.406500	197.790500
Z	−11.107	−29.367	−27.995
Asymp. Sig. (2-tailed)	0.000	0.000	0.000

Table A19. Ranks table of Mann–Whitney U test for specular solar reflectance of WG3 during the three holding time.

		Ranks		
	WG	**N**	**Mean Rank**	**Sum of Ranks**
	WG3T20	607	308.28	187.12700
	WG3T40	607	906.72	550.37800
	Total	1214		
Reflectance	WG3T20	607	335.00	203.34500
	WG3T60	607	880.00	534.16000
	Total	1214		
	WG3T40	607	911.00	552.97700
	WG3T60	607	304.00	184.52800
	Total	1214		

Table A20. Test statistics table of Mann–Whitney U test for specular solar reflectance of WG3 during the three holding time.

	Test Statistics		
	Reflectance		
	WG3T20_WG3T40	**WG3T20_WG3T60**	**WG3T40_WG3T60**
Mann–Whitney U	2599.000	18.817000	0.000
Wilcoxon W	187.127000	203.345000	184.528000
Z	−29.747	−28.977	−32.251
Asymp. Sig. (2-tailed)	0.000	0.000	0.000

References

1. Blengini, G.A.; Busto, M.; Fantoni, M.; Fino, D. Eco-efficient waste glass recycling: Integrated waste management and green product development through LCA. *Waste Manag.* **2012**, *32*, 1000–1008. [CrossRef] [PubMed]
2. Allevi, E.; Gnudi, A.; Konnov, I.V.; Oggioni, G. Municipal solid waste management in circular economy: A sequential optimization model. *Energy Econ.* **2021**, *100*, 105383. [CrossRef]
3. Shao, Y.; Lefort, T.; Moras, S.; Rodriguez, D. Studies on concrete containing ground waste glass. *Cem. Concr. Res.* **2000**, *30*, 91–100. [CrossRef]
4. Conte, S.; Zanelli, C.; Molinari, C.; Guarini, G.; Dondi, M. Glassy wastes as feldspar substitutes in porcelain stoneware tiles: Thermal behaviour and effect on sintering process. *Mater. Chem. Phys.* **2020**, *256*, 123613. [CrossRef]
5. Demir, I. Reuse of waste glass in building brick production. *Waste Manag. Res.* **2009**, *27*, 572–577. [CrossRef]
6. Lassinantti Gualtieri, M.; Mugoni, C.; Guandalini, S.; Cattini, A.; Mazzini, D.; Alboni, C.; Siligardi, C. Glass recycling in the production of low-temperature stoneware tiles. *J. Clean. Prod.* **2018**, *197*, 1531–1539. [CrossRef]
7. Phonphuak, N.; Kanyakam, S.; Chindaprasirt, P. Utilization of waste glass to enhance physical-mechanical properties of fired clay brick. *J. Clean. Prod.* **2016**, *112*, 3057–3062. [CrossRef]
8. Abdeen, H.; Shihada, S. Properties of Fired Clay Bricks Mixed with Waste Glass. *J. Sci. Res. Rep.* **2017**, *13*, 1–9. [CrossRef]

9. Silva, R.V.; de Brito, J.; Lye, C.Q.; Dhir, R.K. The role of glass waste in the production of ceramic-based products and other applications: A review. *J. Clean. Prod.* **2017**, *167*, 346–364. [CrossRef]

10. Kalirajan, M.; Ranjeeth, R.; Vinothan, R.; Vidyavathy, S.M.; Srinivasan, N.R. Influence of glass wastes on the microstructural evolution and crystallization kinetics of glass-ceramic glaze. *Ceram. Int.* **2016**, *42*, 18724–18731. [CrossRef]

11. Dal Bó, M.; Bernardin, A.M.; Hotza, D. Formulation of ceramic engobes with recycled glass using mixture design. *J. Clean. Prod.* **2014**, *69*, 243–249. [CrossRef]

12. Anggono, A.D.; Lopo, E.B.; Sedyono, J.; Riyadi, T.W.B. Fabrication of glaze material from recycled bottle glass and kaolin. *Adv. Sci. Technol. Eng. Syst.* **2019**, *4*, 313–320. [CrossRef]

13. de Souza-Dal Bó, G.C.; Bó, M.D.; Bernardin, A.M. Reuse of laminated glass waste in the manufacture of ceramic frits and glazes. *Mater. Chem. Phys.* **2021**, *257*, 123847. [CrossRef]

14. Gol, F.; Yilmaz, A.; Kacar, E.; Simsek, S.; Sarıtas, Z.G.; Ture, C.; Arslan, M.; Bekmezci, M.; Burhan, H.; Sen, F. Reuse of glass waste in the manufacture of ceramic tableware glazes. *Ceram. Int.* **2021**, *47*, 21031–21068. [CrossRef]

15. Zimmer, A.; Bragança, S.R. A review of waste glass as a raw material for whitewares. *J. Environ. Manag.* **2019**, *244*, 161–171. [CrossRef]

16. Kurtulus, R.; Kavas, T. Investigation on the physical properties, shielding parameters, glass formation ability, and cost analysis for waste soda-lime-silica (SLS) glass containing SrO. *Radiat. Phys. Chem.* **2020**, *176*, 109090. [CrossRef]

17. Kurtulus, R.; Kavas, T.; Akkurt, I.; Gunoglu, K. An experimental study and WinXCom calculations on X-ray photon characteristics of Bi2O3- and Sb2O3-added waste soda-lime-silica glass. *Ceram. Int.* **2020**, *46*, 21120–21127. [CrossRef]

18. Kurtulus, R.; Kavas, T.; Akkurt, I.; Gunoglu, K. Theoretical and experimental gamma-rays attenuation characteristics of waste soda-lime glass doped with La2O3 and Gd2O3. *Ceram. Int.* **2021**, *47*, 8424–8432. [CrossRef]

19. Kurtulus, R.; Kavas, T.; Mahmoud, K.A.; Akkurt, I.; Gunoglu, K.; Sayyed, M.I. Evaluation of gamma-rays attenuation competences for waste soda-lime glass containing MoO3: Experimental study, XCOM computations, and MCNP-5 results. *J. Non-Cryst. Solids.* **2021**, *557*, 120572. [CrossRef]

20. Hu, B.; Hui, W. Lead recovery from waste CRT funnel glass by high-temperature melting process. *J. Hazard. Mater.* **2018**, *343*, 220–226. [CrossRef]

21. Lv, J.; Yang, H.; Jin, Z.; Ma, Z.; Song, Y. Feasibility of lead extraction from waste Cathode-Ray-Tubes (CRT) funnel glass through a lead smelting process. *Waste Manag.* **2016**, *57*, 198–206. [CrossRef] [PubMed]

22. Meng, W.; Wang, X.; Yuan, W.; Wang, J.; Song, G. The Recycling of Leaded Glass in Cathode Ray Tube (CRT). *Procedia Environ. Sci.* **2016**, *31*, 954–960. [CrossRef]

23. Singh, N.; Wang, J.; Li, J. Waste Cathode Rays Tube: An Assessment of Global Demand for Processing. *Procedia Environ. Sci.* **2016**, *31*, 465–474. [CrossRef]

24. Westland, S. CIE Whiteness. In *Encyclopedia of Color Science and Technology*; Luo, R., Ed.; Springer: Berlin/Heidelberg, Geramny, 2015. [CrossRef]

25. Samoilenko, N.; Shchukina, L.; Baranova, A. Development of engobe composition with the use of pharmaceutical glass waste for glazed ceramic granite. *East. Eur. J. Enterp. Technol.* **2021**, *4*, 6–12. [CrossRef]

26. Bloomfield, L. *How Things Work: The Physics of Everyday Life*, 2nd ed.; Wiley: New York, NY, USA, 2001.

Article

The Reuse of Industrial By-Products for the Synthesis of Innovative Porous Materials, with the Aim to Improve Urban Air Quality

Antonella Cornelio [1], Alessandra Zanoletti [1], Roberto Braga [2], Laura Eleonora Depero [1] and Elza Bontempi [1,*]

[1] INSTM and Chemistry for Technologies Laboratory, University of Brescia, via Branze 38, 25123 Brescia, Italy; a.cornelio001@unibs.it (A.C.); alessandra.zanoletti@unibs.it (A.Z.); laura.depero@unibs.it (L.E.D.)

[2] Dipartimento di Scienze Biologiche Geologiche e Ambientali, Università di Bologna, Piazza di Porta San Donato 1, 40126 Bologna, Italy; r.braga@unibo.it

* Correspondence: elza.bontempi@unibs.it

Abstract: This works concerns the characterization and the evaluation of adsorption capability of innovative porous materials synthesized by using alginates and different industrial by-products: silica fume and bottom ash. Hydrogen peroxide was used as pore former to generate a porosity able to trap particulate matter (PM). These new materials are compared with the reference recently proposed porous SUNSPACE hybrid material, which was obtained in a similar process, by using silica fume. Structural, morphological, colorimetric and porosimetric analyses were performed to evaluate the differences between the obtained SUNSPACE typologies. The sustainability of the proposed materials was evaluated in terms of the Embodied Energy and Carbon Footprint to quantify the benefits of industrial by-products reuse. Adsorption tests were also performed to compare the ability of samples to trap PM. For this aim, titania suspension, with particles size about 300 nm, was used to simulate PM in the nanoparticle range. The results show that the material realized with bottom ash has the best performance.

Keywords: air quality; air pollution; SUNSPACE; PM removal; azure chemistry; circular economy; sustainability; SDG 11; SDG 12

Citation: Cornelio, A.; Zanoletti, A.; Braga, R.; Depero, L.E.; Bontempi, E. The Reuse of Industrial By-Products for the Synthesis of Innovative Porous Materials, with the Aim to Improve Urban Air Quality. *Appl. Sci.* **2021**, *11*, 6798. https://doi.org/10.3390/app11156798

Academic Editor: Mónica Calero de Hoces

Received: 2 July 2021
Accepted: 21 July 2021
Published: 23 July 2021

Publisher's Note: MDPI stays neutral with regard to jurisdictional claims in published maps and institutional affiliations.

1. Introduction

Currently, air pollution is one of the main environmental problems that must be faced at the global level. Different studies reveal that more than 90% of humans lives in areas subjected to air pollution [1]. Among the air pollutants, particulate matter (PM) causes the greatest concern [2,3]. PM is generated from natural or anthropogenic sources. The first consists mostly of sea salt, mineral dust (typically from deserts) and volcanic particulate [4] and can be achieved by re-suspended dust or mechanical shearing [5], and the second is derived from combustion processes, vehicle emissions, industrial processing or domestic heating [6]. In particular, a higher concentration of PM occurs during winter seasons, due to car traffic and domestic heating when low temperatures and atmospheric stability hinder PM dispersion [7]. It is estimated that PM is responsible for 8.3 million premature deaths in the world in 2017 [8]. One of the most investigated aspects of PM is its dimension, which strongly determines its capability to penetrate the respiratory system. PM can be classified according to its particle size in PM_{10} ($\leq$10 µm aerodynamic diameter) and $PM_{2.5}$ ($\leq$2.5 µm aerodynamic diameter).

The World Health Organization (WHO) reports that long exposure to $PM_{2.5}$ can generate respiratory and cardiovascular problems [9]. PM is also responsible for diabetes [10], ischemic heart disease and respiratory infection [11]. Moreover, [12] has shown that PM deposition in the respiratory system is higher in the alveolar region followed by the nasal and tracheobronchial region. The International Agency for Research on Cancer revealed that $PM_{2.5}$ is carcinogenic to humans [8].

During the pandemic, PM was considered a possible source of COVID-19 spread [13–16], generating additional concerns for populations living in large cities, which are generally the most polluted in terms of PM concentrations [17].

To reduce its concentration, it is possible to act on PM pollution upstream, by controlling the source emissions, or downstream, by removing the particles by means of chemical methods, dry and wet deposition [18]. Literature reveals that vegetation is one of the most efficient methods for air purification through dry deposition onto leaves [3].

Recently, a new innovative and sustainable porous material called SUNSPACE, "SUstaiNable materials Synthesized from by-Products and Alginates for Clean air and better Environment", has been developed to remove PM from urban areas [19,20]. The importance of this material is twofold: it improves air quality and enables the reuse of industrial by-products.

Economic growth and improved living conditions in the more industrialized countries have led to an increasing demand for goods. This led to large-scale waste production, increasing environmental pollution. The shift from a linear economy to a circular economy marked the turning point. Promoting a more efficient use of resources and recycling, the reuse of wastes allows for the preservation of natural material and reduces landfill disposal [21]. Among the most problematic waste to recycle is industrial waste. It is necessary to ensure that its reuse (for example, after its stabilization) does not create problems for the environment or human health.

In this framework, it is important to highlight that the proposed research activity is in accordance with the SDGs (Sustainable Development Goals), particularly with SDG 11 (Make cities and human settlements inclusive, safe, resilient and sustainable) and SDG 12 (Ensure sustainable consumption and production patterns). Indeed, this work proposes the synthesis and characterization of different SUNSPACE typologies. These new sustainable materials (generated from by-products and waste such as silica fume and bottom ash) are able to increase the resilience of cities [22]. This paper shows an example of circular economy applied in the framework of the Azure Chemistry approach [23].

The original SUNSPACE was synthesized by using silica fume (SF). This by-product is a fine powder (with particle size ranging from 20 to 500 nm) derived from ferro-silicon and silicon metal alloy processing. It is commonly used in construction materials, for example, cement, concrete, bricks or ceramics. It is also demonstrated that SF has good performance as a heavy metals stabilizer [24]. Recently, it has been used in the synthesis of porous material to PM capture [19,20]. The color of silica can change from light to dark grey as a function of the manufacturing process, considering for example, the composition of starting materials, the combustion temperature and the obtained final products [25]. Silica fume used in SUNSPACE synthesis is dark grey. The dark color represents an esthetic limit if the idea is to apply SUNSPACE as plaster on wall or as rooftiles. To overcome this problem, another industrial by-product, bottom ash (BA), was recently proposed for material synthesis and is mainly used with the goal of obtaining a light-colored material [20].

BA is the residue of municipal solid waste incineration (MSWI). It is estimated that BA represents 90% of solid residues generated in incineration plants [21]. Twenty million tons of BA are produced every year in Europe [26]. BA is not considered hazardous waste (2008/98/CE); it contains heavy metal with low volatilization.

Assi et al. revealed that fine fraction of BA (lower than 300 microns) is characterized by a higher leaching concentration of heavy metals, such as Pb and Zn, compared to coarse fraction (more than 1400 microns) [27]. The chemical composition of these ashes (high amount of Al, Si, Ca and Fe) makes them suitable mainly for cement production [28]. Recent work shows the possibility to use BA as a stabilizer agent for fly ash derived from the same incineration process [29]. BA is also used as adsorbent material due to high porosity for heavy metal wastewater treatments [30].

Moreover, another improvement has been introduced in this work. For the first time, hydrogen peroxide was used instead of sodium bicarbonate as a pore former to avoid thermal treatment. This allows for the simplification of the application of materials as plaster and then increases the material sustainability.

Structural, morphological, colorimetric and porosimetric analyses of all SUNSPACE typologies were performed. Additionally, adsorption tests were made to compare the ability of all samples to trap PM.

2. Materials and Methods

2.1. Materials and Reagents

Calcium iodate ($Ca(IO_3)_2$, CAS number: 7789-80-2,) sodium alginate (SA, CAS number: 9005-38-3, viscosity c = 1% water at 25 °C 5.0–40.0 cps), sodium bicarbonate ($NaHCO_3$, CAS number 14455-8, $\geq$99.8% *w/w*) and hydrogen peroxide (H_2O_2, CAS number 7722-84-1, 30% *w/w*) were bought from Sigma Aldrich (St. Louis, Missouri, USA).

Silica fume was kindly provided by Metalleghe spa (Brescia, Italy). Bottom ash, derived from a municipal solid waste incineration plant, and calcium hydroxide were handled by A2A spa. Titanium dioxide powder (Hombitan 97% with 300 nm size) was kindly supplied by Rifra Masterbatches spa (Molinetto di Mazzano, Brescia, Italy).

MilliQ water (Millipore DirectQ-5 purification system) was used for the preparation of porous materials.

2.2. Materials Characterization

Ces Selector software [31] was used to assess the environmental impact of porous materials in terms of the Embodied Energy and Carbon Footprint.

The evaluation of PM entrapment was performed by a total reflection X-ray fluorescence (TXRF) spectrophotometer equipped with Mo anode (S2 PICOFOX, Bruker AXS Microanalysis GmbH (Berlin, Germany)) operating at 750 A and 50 kV. Titania aerosol was generated by Grimm aerosol, Particle-Generator MODEL 7.811.

Structural analysis on porous samples was performed by X-ray-diffraction (XRD) with a Panalytical X'Pert Pro diffractometer equipped with the X'Celerator detector and Cu anode (CuKalpha 1.5406 A) operating at 40 KV and 40 mA. The pattern was collected between 10° and 70° (in 2theta).

The color variation between samples was evaluated using a spectrophotometer (CM-2600d Minolta).

Samples were digested following the EPA method 3052 for siliceous matrices [32] using 4 mL of HNO_3 ($\geq$65%), 2 mL of HCl (37%) and 2 mL of HF (48%). The volume of each digested sample was adjusted to 50 mL using MilliQ water.

To quantify the TiO_2 concentration, an internal standard Ga (1 mg/L) was added to the digested samples. Ten microliters of this solution were deposited onto a plexiglass circular support and dried. The choice to use plexiglass support instead of quartz is due to the HF adoption in digestion process. Three specimens for each sample were performed. Finally, the spectra evaluation was executed by Spectra Plus 5.3 (Bruker AXS Microanalysis GmbH, Berlin, Germany).

The accessible porosity and pore size distribution were quantified using a Pascal 240 Mercury Porosimeter from Thermo Scientific, capable of intruding mercury by a stepwise increase of pressure from vacuum to 200 MPa. After drying at 60 °C for 12 h, a broken piece of each sample of about 1 × 1 cm size was tested. The volume change of mercury between each pressure step gives a number of parameters, of which the pore size distribution, the accessible porosity, the mode, mean and median pore size values were evaluated from the results.

2.3. Samples Preparation

Six samples were synthetized by different industrial by-products (silica fume or bottom ash). The composition of each sample, expressed in g except for hydrogen peroxide expressed in mL, is reported in Table 1.

Table 1. Composition of porous samples. Samples 1, 5 and 6 were synthetized by silica fume, while samples 2, 3 and 4 were synthesized by bottom ash.

	Silica Fume (SF) (g)	Bottom Ash (BA) (g)	Ca(OH)$_2$ (g)	NaC$_6$H$_7$O$_6$ (g)	Ca(IO$_3$)$_2$ (g)	NaHCO$_3$ (g)	H$_2$O$_2$ (mL)
Sample 1	17.88	-	-	0.6	1	5	-
Sample 2	-	9	9	0.6	1	5	-
Sample 3	-	9	9	-	-	-	7.2
Sample 4	-	9	9	-	-	-	5.4
Sample 5	17.88	-	-	0.6	1	-	7.2
Sample 6	17.88	-	-	0.6	1	-	5.4

First, BA was dried for 1 h at 100 °C to eliminate moisture and then sieved to obtain a fine powder with a size dimension lower than 300 μm. Sample 1 and sample 2 were synthetized according to [20]. Sodium alginate and calcium iodate (used as a cross linker) were mixed vigorously with MilliQ water. Then, silica fume (sample 1) or bottom ash and calcium hydroxide, in the same proportion (sample 2), were added to the compound. Finally, sodium bicarbonate (as pore former) was added. The slurry obtained was put on a heating plate for 1 h at a low temperature (70/80 °C) to allow for pore generation due to sodium bicarbonate decomposition.

To avoid the thermal treatment, sodium bicarbonate was replaced with hydrogen peroxide in the synthesis process of samples 3, 4, 5 and 6. Sample 3 was performed by mixing 9 g of bottom ash, 9 g of calcium hydroxide, 7.2 mL of hydrogen peroxide and 9 mL of MilliQ water. The amorphous phase of BA is expected to react with hydrogen peroxide to promote pozzolanic reaction [33], which favors the formation of a stable material. Sample 4 is characterized by a lower quantity of hydrogen peroxide (5.4 mL) compared to sample 3. Sample 5 was synthetized by mixing vigorously 20 mL of MilliQ water, 0.6 g of sodium alginate and 1 g of calcium iodate until a gel is formed. Then, 17.88 g of silica fume and 7.2 mL of hydrogen peroxide were added. Sample 6 contains a lower quantity of hydrogen peroxide (5.4 mL). After the synthesis, all samples were left at room temperature for 1 week to reach the final consolidation. Figure 1 shows the images of all samples after consolidation.

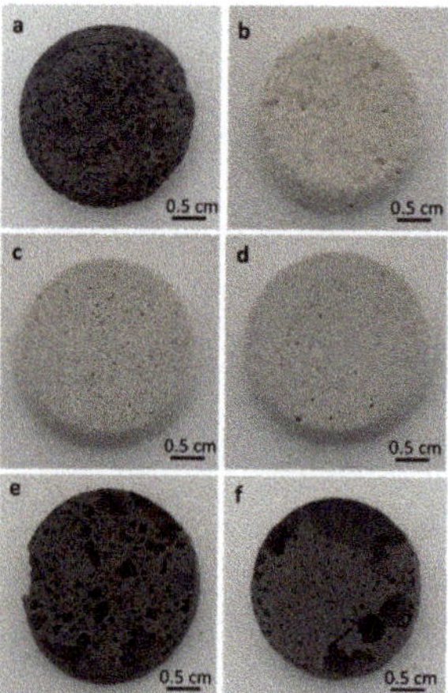

Figure 1. Images of samples synthetized with silica fume (**a,e,f**) and with bottom ash (**b–d**).

3. Results and Discussion

3.1. Sample Sustainability

The sustainability analysis of all samples was evaluated in terms of the Embodied Energy (EE) and Carbon Footprint (CF) [34–37]. The bottom ash pre-treatment and the thermal treatment on a heating plate at a low temperature (70/80 °C), to which some samples were subjected, was considered in the evaluations. In Figure 2, the EE (MJ/kg) versus CF (kg/kg) of six samples is reported. All samples appear to be more sustainable compared to materials commonly used in the filter production such as polypropylene (PP), polyethylene (PE), polyamide (PA) and polystyrene (PS). The samples' sustainability can be compared with building materials such as brick, Portland cement and concrete due to the idea to propose SUNSPACE use in construction. The graph shows that sample 1 has lower EE, while samples 5 and 6 are more sustainable in terms of CO_2 emissions compared to other samples. The higher EE and CF resulting for samples 2, 3 and 4 can be attributed to the use of BA instead of SF in their synthesis. In particular, no pre-treatment is necessary for silica fume before its reuse. Instead, BA was subjected to a thermal treatment, to reduce its moisture, to control the amount of water that was added (see Table 1) and to guarantee the sample's workability during the synthesis. However, in the future, it may also be possible to find more sustainable BA pre-treatment to increase the sustainability of its use.

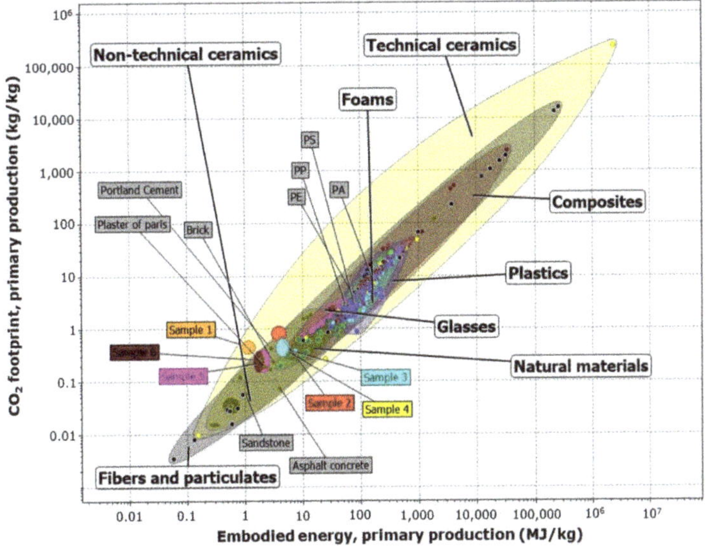

Figure 2. Sustainability analysis of all six samples in terms of the Embodied Energy (EE) and Carbon Footprint (CF).

3.2. Sample Colorimeter

BA is used instead of SF with the aim to lighten the porous material, as shown in Figure 1. Colorimetric analysis allows for the definition of a color with CIELAB coordinates: L (luminance) and two colors channel (a and b). L values are presented between 0 (dark) and 100 (white), while a and b are presented between negative and positive values. Negative values represent green and blue colors, while positive values are red and yellow for a and b, respectively. Table 2 shows the results of colorimetric analysis of all samples. It is evident that the colors are very different. Indeed, samples synthetized with SF (samples 1, 5 and 6) have lower L values (38.83 < L < 47.46) compared with samples obtained starting from BA (samples 2, 3 and 4) (81.97 < L < 86.01).

Table 2. Colorimetric analysis of all samples. Data found including (SCI) and excluding (SCE) specular component.

Samples	SCI			SCE		
	L	a	b	L	a	b
Sample 1	38.83 ± 0.01	−0.92 ± 0.01	−2.33 ± 0.01	39.61 ± 0.01	−0.89 ± 0.02	−2.36 ± 0.01
Sample 2	81.97 ± 0.01	0.77 ± 0.01	6.34 ± 0.01	81.49 ± 0.01	0.80 ± 0.01	6,22 ± 0.01
Sample 3	86.32 ± 0.01	0.16 ± 0.01	3.13 ± 0.01	85.86 ± 0.01	0.21 ± 0.01	3.03 ± 0.02
Sample 4	86.01 ± 0.01	0.21 ± 0.01	3.41 ± 0.01	85.82 ± 0.01	0.26 ± 0.10	3.31 ± 0.01
Sample 5	46.69 ± 0.39	0.63 ± 0.02	0.76 ± 0.24	46.45 ± 0.39	0.59 ± 0.02	0.83 ± 0.24
Sample 6	47.46 ± 0.52	0.81 ± 0.02	1.46 ± 0.06	47.21 ± 0.51	0.78 ± 0.01	1.53 ± 0.07

3.3. Structural Analysis

An XRD analysis of samples 1 (SUNSPACE) and 2 (SUNSPACE with BA) is reported in previous work [20]. In Figure 3, the XRD patterns of samples 3, 4, 5 and 6 are shown. Samples 3 and 4 show similar patterns. The identified crystalline phases are calcium carbonate ($CaCO_3$) and quartz (SiO_2), both predominant in bottom ash powder [29], and calcium hydroxide ($Ca(OH)_2$) used in the material synthesis. Moreover, calcium carbonate is also probably formed during the carbonation reaction between calcium hydroxide and CO_2, as already extensively discussed in literature [38–41]. Some peaks can be attributed to aluminum silicate hydroxide chloride. From this analysis, it is possible to conclude that the hydrogen peroxide concentration does not seem to involve structural changes in the samples. Additionally, samples 5 and 6 show similar XRD patterns. They are characterized by the presence of a large halo (between 15° and 30° in 2theta), that can be attributed to an amorphous phase due to the presence of silica fume, which is an amorphous material [17,24,42,43]. Furthermore, cristobalite (SiO_2) and lautarite ($Ca(IO_3)_2$) peaks are identified. Calcium iodate is used in the synthesis of these samples.

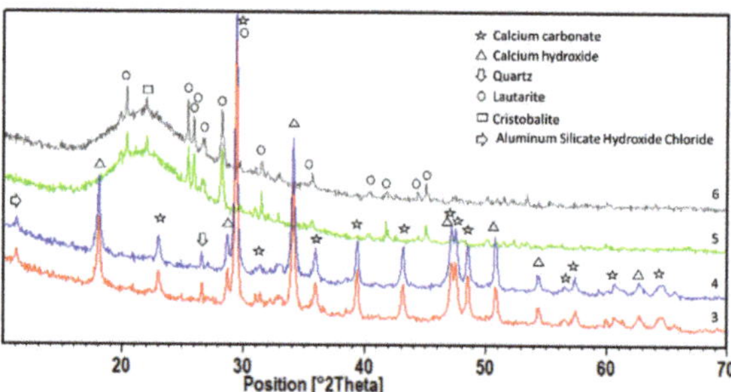

Figure 3. XRD patterns of samples 3, 4, 5 and 6.

3.4. Morphological Analysis

Figure 4 shows the optical microscopy of samples 1 (a), 2 (b), 3 (c), 4 (d), 5 (e) and 6 (f). All samples show a visible porosity. The main differences concern the shape and size of the pores. Sample 1 reveals the presence of macro-porous between 100 and 200 μm, while sample 2 is characterized by pores with sizes up to 1 mm. Sample 3 shows a highly porous surface, rich in circular pores. The larger pores have a diameter of about 300 μm. The morphology of sample 4 is similar to that of the previous one, but there are some larger pores. This is probably correlated to the lower amount of pore former used for the synthesis of this sample. Samples with SF present pores with irregular shapes and dimensions of about 100–200 μm (sample 5) and 500 μm (sample 6).

3.5. Porosimetric Analysis

The pore size distributions of samples are shown in Figure 5, while Table 3 summarizes the basic descriptive statistics.

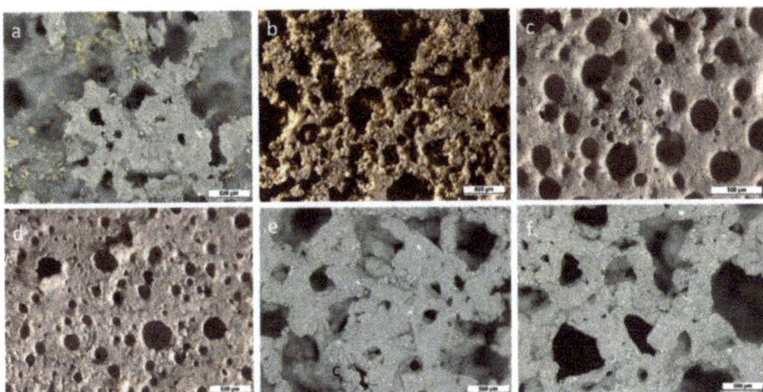

Figure 4. Optical microscopy images of samples sythetized with silica fume: sample 1 (**a**), 5 (**e**) and 6 (**f**) and bottom ash: sample 2 (**b**), sample 3 (**c**) and 4 (**d**).

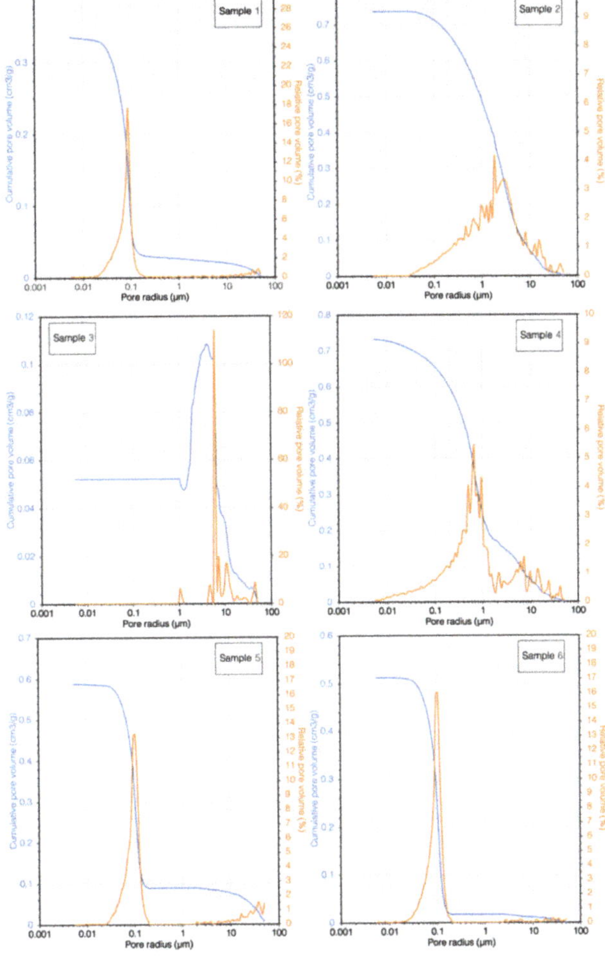

Figure 5. Pore size distributions of all six samples quantified by mercury porosimetry analysis.

Table 3. Results from mercury intrusion porosimetry.

Sample	Accessible Porosity (%)	Average Pore Radius (μm)	Median Pore Radius (μm)	Modal Pore Radius (μm)
1	41.1	0.0356	0.0777	0.0849
2	63.2	0.0282	1.7808	1.7938
3	3.4	0.0141	9.6896	5.6504
4	58.1	0.0387	0.6153	0.6127
5	51.6	0.0418	0.0954	0.1058
6	49.1	0.0455	0.0911	0.1063

The accessible porosity by intrusion is always above 41%, with a maximum of 63% for the sample 2. The anomalously low value for sample 3 (about 3% accessible porosity), which is clearly at odds with the morphological observations, indicates sample deformation and porosity collapse during the test. The sample 3 is therefore not considered in the mercury porosimetry results.

The pore sizes of the samples 1, 5 and 6, all of which were prepared with SF, follow a unimodal distribution with a pore radius mode of 0.08–0.1 μm. The sample 4 shows a bimodal pore size distribution, with a main peak at about 0.65 μm and a secondary peak related to larger pores with a radius of about 7–8 μm. Finally, the pore size distribution of sample 2 spans about an order of magnitude, with a mode of 1.79 μm.

3.6. Sample PM Capability

The performance in PM trapping of SUNSPACE and SUNSPACE synthesized by using BA was evaluated in previous work [20]. In this paper, these values were compared with samples 3, 4, 5 and 6.

To evaluate the adsorption capacity of these porous materials, three samples of each material were exposed to a TiO_2 suspension with size dimension of 300 nm, 3 g/L. The suspension was prepared according to [20] and then sprayed using Grimm aerosol nanoparticle generator. The same experimental set up of [20] was performed. Samples were exposed for 4 min, after 1 min of stabilization.

To evaluate the amount of trapped TiO_2 by each specimen, pristine and exposed samples were compared. They were scraped to obtain 0.25 g of powder from each sample for chemical analysis. For this aim, samples were digested according to the procedure reported in the experimental section. The Ti concentration (mg/L) was evaluated by TXRF analysis and then converted to TiO_2 expressed in mg/kg.

Figure 6 shows the comparison between pristine and exposed samples synthetized by BA (Figure 6a) (sample 3 and 4) and SF (Figure 6b) (sample 5 and 6). The results show that pristine materials always present a titania amount lower than that of corresponding materials, which were subjected to titania flux.

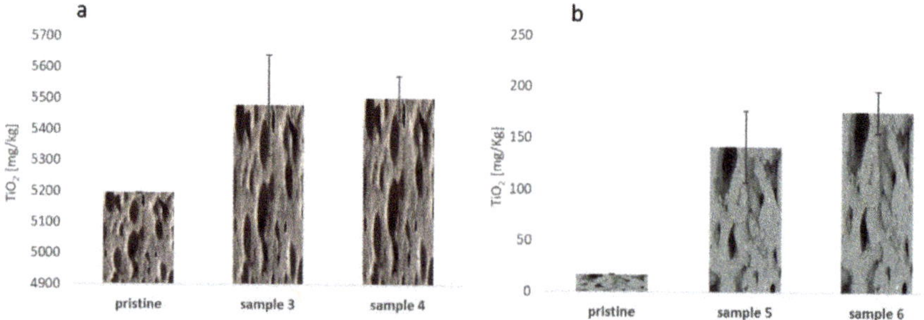

Figure 6. Amount of TiO_2 concentration on SUNSPACE samples synthetized by (**a**) bottom ash (samples 3 and 4) and (**b**) silica fume (samples 5 and 6). Pristine samples were not in contact with titania nanoparticles. The other samples were exposed to TiO_2.

It is also apparent that the pristine samples synthetized with BA are characterized by a higher concentration of titania than those generated with SF. Indeed, as reported in the literature, BA is characterized by titania (about 1%wt) [27,28,44,45].

Using both using BA and SF, samples 4 and 6 (characterized by a lower hydrogen peroxide concentration) obtained good results despite their different pore structure. Sample 4 is characterized by a bimodal pore size distribution. The presence of interconnected pores with sizes more than 1 μm suggests an open structure that favors the uptake of PM. Sample 6 shows a uniform structure, with pores of a lower dimension (about 0.1 μm). The graph shows that sample 4 seems to have the best performance. High concentrations of TiO_2 (about 300 mg/kg) penetrated into its porous surface.

4. Conclusions

This work shows a comparison of innovative porous materials synthetized by bottom ash and silica fume (as raw materials) to promote circular economy and new solutions to improve urban air quality. Some variations in the synthesis process were performed in the framework of the Azure chemistry approach. The use of hydrogen peroxide instead of sodium bicarbonate circumvented the heating process. This simplifies the synthesis process and, consequently, the application of material as plaster. All of the samples were characterized by structural, morphological, porosimetric and colorimetric analyses. The use of BA results in a lighter-colored material than the one realized with SF. Samples realized by SF and BA show different porous structures. The materials synthetized by SF are characterized by smaller pore sizes (0.08–0.1 μm) than those made with BA (0.65–8 μm). The porous materials investigated in this work were designed to reduce PM concentrations, so an adsorption capacity test was carried out. According to the results, sample 4 (synthetized with bottom ash) shows the best performance. For the future, some variation in the BA treatment will be considered to promote the sustainability of the material.

Author Contributions: Conceptualization, A.C., A.Z. and E.B.; software, A.Z.; formal analysis, A.C. and A.Z.; investigation, A.C., A.Z. and R.B.; data curation, A.C. and A.Z.; writing—original draft preparation, A.C., A.Z., R.B. and E.B.; writing—review and editing, L.E.D. and E.B.; supervision, E.B.; project administration, L.E.D. and E.B.; funding acquisition, L.E.D. and E.B. All authors have read and agreed to the published version of the manuscript.

Funding: This research was conducted in the framework of the RENDERING project: "Energy recovery of waste sludge and their re-use as an alternative to some natural resources, for the production of Green composites", funded by Ministero dell'Ambiente e della Tutela del Territorio e del Mare-Direzione generale dei rifiuti e dell'inquinamento. It is supported by the University of Brescia, CSMT, INSTM and Regione Lombardia.

Institutional Review Board Statement: Not applicable.

Informed Consent Statement: Not applicable.

Data Availability Statement: Data is contained within the article.

Acknowledgments: Fausto Peddis (Dip. Di Ingegneria Civile, Chimica, Ambientale e dei materiali—DICAM).

Conflicts of Interest: The authors declare no conflict of interest.

References

1. Wei, Z.; Su, Q.; Yang, J.; Zhang, G.; Long, S.; Wang, X. High-performance filter membrane composed of oxidized Poly (arylene sulfide sulfone) nanofibers for the high-efficiency air filtration. *J. Hazard. Mater.* **2021**, *417*, 126033. [CrossRef] [PubMed]
2. Domingo, J.L.; Rovira, J. Effects of air pollutants on the transmission and severity of respiratory viral infections. *Environ. Res.* **2020**, *187*, 109650. [CrossRef]
3. Przybysz, A.; Nersisyan, G.; Gawroński, S.W. Removal of particulate matter and trace elements from ambient air by urban greenery in the winter season. *Environ. Sci. Pollut. Res.* **2019**, *26*, 473–482. [CrossRef]
4. Gieré, R.; Querol, X. Solid particulate matter in the atmosphere. *Elements* **2010**, *6*, 215–222. [CrossRef]

5.   Zanoletti, A.; Bilo, F.; Federici, S.; Borgese, L.; Depero, L.E.; Ponti, J.; Valsesia, A.; La Spina, R.; Segata, M.; Montini, T.; et al. The first material made for air pollution control able to sequestrate fine and ultrafine air particulate matter. *Sustain. Cities Soc.* **2020**, *53*, 101961. [CrossRef]

6.   Juda-Rezler, K.; Reizer, M.; Oudinet, J.-P. Determination and analysis of PM10 source apportionment during episodes of air pollution in Central Eastern European urban areas: The case of wintertime. *Atmos. Environ.* **2011**, *45*, 6557–6566. [CrossRef]

7.   Majewski, G.; Kleniewska, M.; Brandyk, A. Seasonal variation of particulate matter mass concentration and content of metals. *Pol. J. Environ. Stud.* **2011**, *20*, 417–427.

8.   Wu, X.; Zhu, B.; Zhou, J.; Bi, Y.; Xu, S.; Zhou, B. The epidemiological trends in the burden of lung cancer attributable to PM2.5 exposure in China. *BMC Public Health* **2021**, *21*, 1–8. [CrossRef] [PubMed]

9.   Cui, J.; Wang, Y.; Lu, T.; Liu, K.; Huang, C. High performance, environmentally friendly and sustainable nanofiber membrane filter for removal of particulate matter 1.0. *J. Colloid Interface Sci.* **2021**, *597*, 48–55. [CrossRef]

10.   Heck, T.G.; Fiorin, P.B.G.; Frizzo, M.N.; Ludwig, M.S. Fine particulate matter (PM2.5) air pollution and type 2 diabetes mellitus (T2DM): When experimental data explains epidemiological facts. *Diabetes Its Complicat.* **2018**. [CrossRef]

11.   Apte, J.S.; Brauer, M.; Cohen, A.J.; Ezzati, M.; Pope, I.C.A. Ambient PM2.5 reduces global and regional life expectancy. *Environ. Sci. Technol. Lett.* **2018**, *5*, 546–551. [CrossRef]

12.   Lin, Y.; Bahreini, R.; Lee, S.-B.; Bae, G.-N.; Jung, H. Correlations of PM metrics with human respiratory system deposited PM mass determined from ambient particle size distributions and effective densities. *Aerosol Sci. Technol.* **2019**, *54*, 262–276. [CrossRef]

13.   Bontempi, E. First data analysis about possible COVID-19 virus airborne diffusion due to air particulate matter (PM): The case of Lombardy (Italy). *Environ. Res.* **2020**, *186*, 109639. [CrossRef] [PubMed]

14.   Bontempi, E. The Europe second wave of COVID-19 infection and the Italy "strange" situation. *Environ. Res.* **2021**, *193*, 110476. [CrossRef]

15.   Bontempi, E.; Vergalli, S.; Squazzoni, F. Understanding COVID-19 diffusion requires an interdisciplinary, multi-dimensional approach. *Environ. Res.* **2020**, *188*, 109814. [CrossRef] [PubMed]

16.   Bontempi, E. Commercial exchanges instead of air pollution as possible origin of COVID-19 initial diffusion phase in Italy: More efforts are necessary to address interdisciplinary research. *Environ. Res.* **2020**, *188*, 109775. [CrossRef]

17.   Anand, U.; Cabreros, C.; Mal, J.; Ballesteros, F.; Sillanpää, M.; Tripathi, V.; Bontempi, E. Novel coronavirus disease 2019 (COVID-19) pandemic: From transmission to control with an interdisciplinary vision. *Environ. Res.* **2021**, *197*, 111126. [CrossRef] [PubMed]

18.   Shabnam, N.; Oh, J.; Park, S.; Kim, H. Impact of particulate matter on primary leaves of *Vigna radiata* (L.) R. Wilczek. *Ecotoxicol. Environ. Saf.* **2021**, *212*, 111965. [CrossRef] [PubMed]

19.   Zanoletti, A.; Bilo, F.; Borgese, L.; Depero, L.E.; Fahimi, A.; Ponti, J.; Valsesia, A.; La Spina, R.; Montini, T.; Bontempi, E. SUNSPACE, A porous material to reduce air particulate matter (PM). *Front. Chem.* **2018**, *6*, 534. [CrossRef]

20.   Cornelio, A.; Zanoletti, A.; Federici, S.; Depero, L.E.; Bontempi, E. Porous materials derived from industrial by-products for titanium dioxide nanoparticles capture. *Appl. Sci.* **2020**, *10*, 8086. [CrossRef]

21.   Kasina, M.; Kajdas, B.; Michalik, M. The leaching potential of sewage sludge and municipal waste incineration ashes in terms of landfill safety and potential reuse. *Sci. Total. Environ.* **2021**, *791*, 148313. [CrossRef]

22.   Bontempi, E.; Sorrentino, G.; Zanoletti, A.; Alessandri, I.; Depero, L.; Caneschi, A. Sustainable materials and their contribution to the sustainable development goals (SDGs): A critical review based on an Italian example. *Molecules* **2021**, *26*, 1407. [CrossRef]

23.   Zanoletti, A.; Bilo, F.; Depero, L.E.; Zappa, D.; Bontempi, E. The first sustainable material designed for air particulate matter capture: An introduction to Azure Chemistry. *J. Environ. Manag.* **2018**, *218*, 355–362. [CrossRef]

24.   Rodella, N.; Bosio, A.; Dalipi, R.; Zacco, A.; Borgese, L.; Depero, L.E.; Bontempi, E. Waste silica sources as heavy metal stabilizers for municipal solid waste incineration fly ash. *Arab. J. Chem.* **2017**, *10*, S3676–S3681. [CrossRef]

25.   Panjehpour, M.; Abdullah, A.; Ali, A.; Demirboga, R. A review for characterization of silica fume and its effects on concrete properties. *Int. J. Sustain. Constr. Eng. Technol.* **2011**, *2*, 1–7.

26.   Šyc, M.; Simon, F.G.; Hykš, J.; Braga, R.; Biganzoli, L.; Costa, G.; Funari, V.; Grosso, M. Metal recovery from incineration bottom ash: State-of-the-art and recent developments. *J. Hazard. Mater.* **2020**, *393*, 122433. [CrossRef]

27.   Assi, A.; Bilo, F.; Federici, S.; Zacco, A.; Depero, L.E.; Bontempi, E. Bottom ash derived from municipal solid waste and sewage sludge co-incineration: First results about characterization and reuse. *Waste Manag.* **2020**, *116*, 147–156. [CrossRef]

28.   Clavier, K.A.; Paris, J.M.; Ferraro, C.C.; Bueno, E.T.; Tibbetts, C.M.; Townsend, T.G. Washed waste incineration bottom ash as a raw ingredient in cement production: Implications for lab-scale clinker behavior. *Resour. Conserv. Recycl.* **2021**, *169*, 105513. [CrossRef]

29.   Assi, A.; Bilo, F.; Zanoletti, A.; Ponti, J.; Valsesia, A.; La Spina, R.; Zacco, A.; Bontempi, E. Zero-waste approach in municipal solid waste incineration: Reuse of bottom ash to stabilize fly ash. *J. Clean. Prod.* **2020**, *245*, 118779. [CrossRef]

30.   Zhang, H.Y.; Zheng, Y.; Hu, H.T.; Qi, J.Y. Use of municipal solid waste incineration bottom ash in adsorption of heavy metals. *Key Eng. Mater.* **2011**, *474-476*, 1099–1102. [CrossRef]

31.   *Granta Design Cambridge Engineering Selector (CES) Software 2019*; Granta Design Ltd.: Cambridge, UK, 2019.

32.   *Us-Epa Method 3052-Microwave Assisted Acid Digestion of Siliceous and Organically Based Matrices*; United States Environmental Protection Agency: Washington, DC, USA, 1996.

33. Kasaniya, M.; Thomas, M.D.; Moffatt, E.G. Pozzolanic reactivity of natural pozzolans, ground glasses and coal bottom ashes and implication of their incorporation on the chloride permeability of concrete. *Cem. Concr. Res.* **2021**, *139*, 106259. [CrossRef]
34. Bontempi, E. A new approach for evaluating the sustainability of raw materials substitution based on embodied energy and the CO_2 footprint. *J. Clean. Prod.* **2017**, *162*, 162–169. [CrossRef]
35. Fahimi, A.; Federici, S.; Depero, L.E.; Valentim, B.; Vassura, I.; Ceruti, F.; Cutaia, L.; Bontempi, E. Evaluation of the sustainability of technologies to recover phosphorus from sewage sludge ash based on embodied energy and CO_2 footprint. *J. Clean. Prod.* **2021**, *289*, 125762. [CrossRef]
36. Ducoli, S.; Zacco, A.; Bontempi, E. Incineration of sewage sludge and recovery of residue ash as building material: A valuable option as a consequence of the COVID-19 pandemic. *J. Environ. Manag.* **2021**, *282*, 111966. [CrossRef]
37. Bontempi, E. *Raw Materials Substitution Sustainability*; Springer Briefs in Applied Science and Technology; Springer International Publishing: Berlin/Heidelberg, Germany, 2017.
38. Benassi, L.; Pasquali, M.; Zanoletti, A.; Dalipi, R.; Borgese, L.; Depero, L.E.; Vassura, I.; Quina, M.; Bontempi, E. Chemical stabilization of municipal solid waste incineration fly ash without any commercial chemicals: First pilot-plant scaling up. *ACS Sustain. Chem. Eng.* **2016**, *4*, 5561–5569. [CrossRef]
39. Assi, A.; Bilo, F.; Zanoletti, A.; Ducoli, S.; Ramorino, G.; Gobetti, A.; Zacco, A.; Federici, S.; Depero, L.E.; Bontempi, E. A Circular Economy Virtuous Example—Use of a Stabilized Waste Material Instead of Calcite to Produce Sustainable Composites. *Appl. Sci.* **2020**, *10*, 754. [CrossRef]
40. Bosio, A.; Rodella, N.; Gianoncelli, A.; Zacco, A.; Borgese, L.; Depero, L.E.; Bingham, P.; Bontempi, E. A new method to inertize incinerator toxic fly ash with silica from rice husk ash. *Environ. Chem. Lett.* **2013**, *11*, 329–333. [CrossRef]
41. Assi, A.; Federici, S.; Bilo, F.; Zacco, A.; Depero, L.E.; Bontempi, E. Increased sustainability of carbon dioxide mineral sequestration by a technology involving fly ash stabilization. *Materials* **2019**, *12*, 2714. [CrossRef]
42. Zanoletti, A.; Vassura, I.; Venturini, E.; Monai, M.; Montini, T.; Federici, S.; Zacco, A.; Treccani, L.; Bontempi, E. A new porous hybrid material derived from silica fume and alginate for sustainable pollutants reduction. *Front. Chem.* **2018**, *6*, 60. [CrossRef]
43. Assi, A.; Bilo, F.; Zanoletti, A.; Ponti, J.; Valsesia, A.; La Spina, R.; Depero, L.E.; Bontempi, E. Review of the reuse possibilities concerning ash residues from thermal process in a medium-sized urban system in northern Italy. *Sustainability* **2020**, *12*, 4193. [CrossRef]
44. Obe, R.K.D.; de Brito, J.; Lynn, C.J.; Silva, R.V. *Sustainable Construction Materials: Municipal Incinerated Bottom Ash*; Elsevier Ltd.: Amsterdam, The Netherlands, 2018.
45. Cheriaf, M.; Cavalante Rocha, J.; Péra, J. Pozzolanic properties of pulverized coal combustion bottom ash. *Cem. Concr. Res.* **1999**, *29*, 1387–1391. [CrossRef]

applied
sciences

Article

Bridging Green Gaps: The Buying Intention of Energy Efficient Home Appliances and Moderation of Green Self-Identity

Ya Li [1,2], Abu Bakkar Siddik [3,*], Mohammad Masukujjaman [4] and Xiujian Wei [1]

1   School of Economics and Finance, Xi'an Jiaotong University, Xi'an 710049, China; wstliya@126.com (Y.L.); weixj@mail.xjtu.edu.cn (X.W.)
2   Medical Management Service Center of Shandong Health Commission (International Health Exchange Center of Shandong Province), Lixia District, Jinan 250014, China
3   School of Economics and Management, Shaanxi University of Science and Technology (SUST), Weiyang District, Xi'an 710021, China
4   Department of Business Administration, Northern University Bangladesh, Banani C/A, Dhaka 1213, Bangladesh; masukujjaman@nub.ac.bd
*   Correspondence: ls190309@sust.edu.cn; Tel.: +86-156-8601-2117

Abstract: This study investigates the factors influencing the buying intention of energy-efficient home appliances in Bangladesh. It also develops a conceptual framework that integrates additional constructs with the theory of planned behavior (TPB) and borrows questions from past literature. Employing a convenience sampling technique, a total of 365 completed structured questionnaires were gathered from various super shops in Dhaka, Bangladesh. The structural equation modeling (SEM) technique was thereafter used to analyze the data with the AMOS 21. The study established that environmental knowledge, attitude, subjective norms, and perceived behavioral control significantly affected the consumers' buying intention of energy-efficient home appliances (EEHA). The result revealed a significant relationship between environmental concern, environmental knowledge, subjective norms, eco-labeling, and attitude towards buying. It also confirmed that the green self-identity moderates the existent relationship between the attitude and buying intention of energy-efficient home appliances, while environmental knowledge does not. The research advances numerous policy suggestions to managers or marketers, as well as future research directions.

Keywords: buying intention; energy-efficient home appliances; green self-identity; theory of planned behavior

Citation: Li, Y.; Siddik, A.B.; Masukujjaman, M.; Wei, X. Bridging Green Gaps: The Buying Intention of Energy Efficient Home Appliances and Moderation of Green Self-Identity. *Appl. Sci.* **2021**, *11*, 9878. https://doi.org/10.3390/app11219878

Academic Editor: Pietro Picuno

Received: 17 September 2021
Accepted: 19 October 2021
Published: 22 October 2021

Publisher's Note: MDPI stays neutral with regard to jurisdictional claims in published maps and institutional affiliations.

1. Introduction

A household sector plays a significant role in energy conservation and environmental sustainability in using energy-saving goods [1]. Smaller energy demand and improved efficiency are widely recognized as the most optimistic, fast, economical, and secure alternative to alleviating environmental deterioration and climate change [2,3]. Globally, policymakers have emphasized the need for individuals to take responsibility for their immediate environment. These include recycling, use of green label products, and reduction in power usage [4]. Products that swiftly dissipate energy are considered important in achieving efficiency and reducing carbon emissions [5–7]. However, energy saves in the residential sector are dependent on users' technological and habitual behavior. Technological choices include consumers' preference for energy-efficient or traditional equipment, while habitual acts include their practices of turning off appliances when not in use [8]. However, energy-efficient appliances (EEA) offer more energy efficiency and sustainability than habitual corrections. Also, they do not require continuous customer efforts and is a one-time expenditure [9].

Energy-efficient appliances use less electricity while maintaining the same performance, convenience, and comfort [10]. Concerns about rising emissions have resulted in investments in energy-efficient & conservation (EEC) projects [11]. However, it is worth noting that the terms "energy efficiency" and "energy conservation" have fundamentally distinct meanings. Energy efficiency entails the minimization of direct energy use through advancements in technology [12], while energy conservation involves the use of technology or modification of human behavior towards the conservation of energy [13]. Consumer behavior must be altered for the successful adoption of EEA, but not for energy conservation [14].

Intensive research has been conducted on the buying behavior of energy-efficient appliances in advanced economies like the UK, Netherlands, Australia, USA, Germany, and Switzerland [4,15–19], and emerging economies like Malaysia, China, Vietnam, India, and South Korea [5,20–23]. From the understanding of the researcher, no similar study has been conducted from the Bangladeshi perspective. Since each country is unique in its socio-economic conditions and resource availability, decision-making on technology choices varies. As a result, an investigation into the barriers and reasons for the lack of the diffusion of energy-saving technology from the Bangladeshi perspective is necessary. As Bangladesh is characterized by a progressive GDP and per capita income growth, it is important to determine what people think of modern and energy-efficient home appliances from a newly developing economic perspective.

The TPB has been utilized in many pro-environmental research areas, including low-consumption appliances [1,5,24–27], and despite its widespread acceptance, the model has been criticized. The major complaint is stemmed on the need for more variables to improve its predictive and explanatory power [6,28,29]. Some researchers claimed that the TPB paradigm fails to address a sufficient variety of intentions [30,31], and as a result, additional factors can be included in the TPB if they provide a significant understanding of the behavior [32,33]. Therefore, to improve the explanatory power of the TPB, researchers have suggested the addition of new variables that are meaningful as much as to theoretically influence intentions [34–36]. In addition, attitude-intention, also regarded as the green gap, is one of the inherited problems of the TPB model. Researchers [9,37–39] studied the attitude-intention gaps and suggested constructs such as ethical consumption behavior, environmental knowledge, and environmental concern, but ignored the construct of green emotion or green responsibility (green self-identity) narrowing the gaps. Thus, there is a critical need to explore these factors empirically.

Previous studies on environmental psychology examined how self-identities predict behavior. For instance, the energy-saving self-identity predicts energy-saving intention [40], while the recycling self-identity predicts recycling behaviors [41]. The same cannot be said for other pro-environmental activities. A broad spectrum of pro-environmental attitudes, intentions, and behaviors are related to a more general environmental self-identity [42,43]. An individual's general green identity influences their pro-environmental action, and as a result, addressing this overall environmental self-identity becomes more useful to spur pro-environmental actions. Qasim et al. [44] used environmental self-identity to moderate the relationship between consumption value and behavioral intent, while Neves and Oliveira [45] used it to moderate the relationship between co-benefits, labeling, operations and maintenance, and savings. However, to the researcher's knowledge, none of these factors assessed the green self-identity as a moderator between attitude-intention to bridge the TPB's inherited gaps. Hence, the scope of this study lies in the identification of the impact of green self-identity on the consumers' buying intention of products like energy-efficient home appliances.

In previous studies [46,47], environmental knowledge was considered as one of the most direct indicators of environmental behavior. However, much research has highlighted a minimal direct association between environmental knowledge and environmental performance [48]. According to Kollmuss and Agyeman [49], the direct relationship existent between green awareness and PEB is still unknown. Since a contradictory result exists in the past literature, there is a need to retest the variable for validation from a developing country's perspective. Likewise, Kim et al. [50] observed a coherent SN-Attitude relationship but not with the BI, while Anssi Tarkiainen Sanna Sundqvist [51] observed a relationship with intention but not with Attitude. However, there is still an unexplored question of whether subjective norms simultaneously affect attitude and buying intention or not.

Therefore, to adequately address the gaps above in the literature, the present study investigates the determinants of Bangladeshis' buying intention of the EEHA. This paper adds to the body of knowledge twofold; a new context and new variables. The present study would like to see the necessary factors in the purchase intention of Bangladeshi shoppers, which will broaden the view of EEHA buying in a developing country context. Regarding the new variables, this study adopts the TPB model while incorporating additional constructs such as green self-identity, environmental concern, environmental knowledge, and eco-labeling. The green self-identity and environmental knowledge and green self-identity were proposed as moderating variables between attitude and buying intention to narrow down the prevailing green gap and provide a comprehensive understanding of the problem.

The remainder of this article is structured as follows: Section 2 offers a review of the relevant literature and formulates hypotheses. Section 3 details the methodology and data collection procedures employed in this investigation. Section 4 discusses the analysis of data to verify the validity and reliability of the employed method and assesses the predicted results produced. Section 5 presents the results and implications of the study. Section 6 addresses the limitations and provides suggestions for future study.

2. Literature Review and Hypothesis Development

The study adopted the theory of planned behavior as the underpinning theory and also added some context-wise factors (environmental concern, environmental knowledge, eco-labeling, and green self-identity) based on the massive literature review (Table 1) in the proposed model to fulfill objectives, as can be seen in Figure 1.

2.1. Theory of Planned Behavior

The theory of planned behavior (TPB) is a common theory employed in the prediction of human behavior that can be easily controlled when compared to other cognitive elements [32]. Owing to its accuracy, it remains one of the most important social psychology theories for the prediction of the conduct of individuals [32] and was first developed as an extension of the reasoned theory of action (TRA). According to the TRA, the customer's use of a product is contingent on their intention to use it, which is created in accordance with social norms [52]. Since the TPB provides a framework for studying behavioral predictors, individual behavior is determined by behavioral intents and influenced by attitudes toward behavior, subjective norms, and perceived behavioral control [53,54]. The TPB states that: "The more the behavioral intentions, the greater the probability of exhibiting a specific behavior.

2.1.1. Attitude and Buying Intention

Attitude refers to a predisposition or proclivity toward a notion, person, or situation. It denotes a broad assessment of an individual's behavior that results in a positive end. The TPB asserts that attitude determines intention, which in turn defines human behavior. As an instance, Yadav and Pathak [55] examined the factors influencing green purchasing behavior and discovered that attitude has a favorable effect on buying intentions. Also, numerous studies have reported that attitude substantially influences the purchase intention of consumers with regards to energy-saving equipment [5,56,57]. Owing to the favorable relationship existent between attitude and intentions towards the purchase of energy-efficient appliances, the following hypothesis is proposed:

Hypothesis 1. *Attitude is significantly and positively related to the intention of residents to purchase EEHA.*

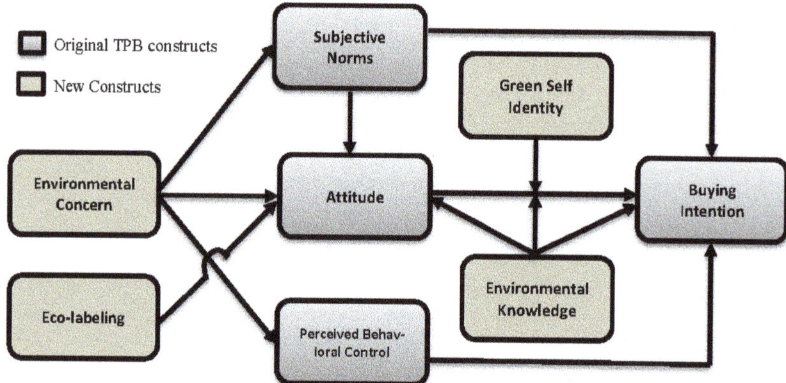

Figure 1. Conceptual Framework [Source: Authors compilation].

2.1.2. Subjective Norms

Subjective norm refers to the perceived social consequences of "performing" or "refraining from performing" [32]. Individuals frequently imitate others and adhere to various social norms. These practices are a society's 'guideline' norms or behavioral desires that govern what is considered acceptable and desirable [57]. Earlier research established the perceived norm as a strong predictor of plans to acquire energy-efficient appliances [58,59]. The latest evidence in South Korea, depending on survey questions to evaluate Korean customers' buying intention toward EEAs, discovered that SN had a favorable effect on EEHA [9]. Furthermore, Wang et al. [60] noted that this impact also applies to EEA purchase intentions. Additional research from around Asia [61,62] has backed up these outcomes. Furthermore, Ali et al. [1] discovered a non-significant connection between SN and purchase intent of energy-saving items in the case of Pakistani consumers' inclination to buy energy-saving products. Nevertheless, Tan et al. [5] identified that SN does not have a substantially positive connection with EEA buying intention, and the absence of a link implies that customers are not easily affected by other people's judgments. As a result, the following hypotheses are proposed:

Hypothesis 2. *Subjective norm is significantly and positively related to the attitude of residents towards the purchase of EEHA.*

Hypothesis 3. *Subjective norm is significantly and positively related to the intention of residents to purchase EEHA.*

2.1.3. Perceived Behavioral Control (PBC)

PBC is a term that relates to the perceived ease or difficulty of engaging in an activity. It is formed by external and internal factors that either encourage or discourage a person from indulging in the action. As an instance, when customers adopt a new product, they establish the effort and resource needs [63]. Convenience factors such as price incentives, appliance availability, and the trustworthiness of energy efficiency labels all play a role in the context of EEHA [64]. Wang et al. [6] evaluated the behavioral intention for electric vehicles by quantifying the PBC in relation to price and availability. The PBC was observed to have a favorable influence on the adoption intentions in this study. The PBC was described by Wang et al. [65] as the perceived discomfort and financial benefits related to an energy conservation activity. Wang et al. [6] also discovered a significant influence of the PBC on energy-saving behavior, resulting in the postulation of the following hypothesis:

Hypothesis 4. *The PBC is significantly and positively related to the intention of residents to purchase EEHA.*

Table 1. The Buying Intention of Energy-Efficient Appliances in Asian Countries.

Country	Authors	Sample/Analysis Method	Constructs Found Significant
Pakistan	[66]	673/PLS-SEM/MGA	Warm glow benefits, utilitarian environmental benefits, perceived behavioral control, normative beliefs, eco-literacy, attitude, and subjective norm
Pakistan	[1]	289/PLS SEM	Attitude, policy information campaigns, past purchase experience, and PBC
China	[67]	1472/Logistic regression/SPSS	Incomes, household size, and dwelling areas
China	[24]	305/PLS SEM	Environmental concern, subjective norm, attitude, environmental knowledge, and PBC
China	[68]	477/SEM/AMOS	Personal norm, PBC, awareness of consequences, attitude, ascription of responsibility, and subjective norm.
India	[69]	300/Multiple Regression/SPSS	Perceived product risk, skepticism towards label claims, price sensitivity, perceived personal inconvenience, and subjective norms
South-Korea	[70]	304/Logistic regression/SPSS	Price and eco-label
Pakistan	[1]	396/PLS SEM	Innovativeness, PBC, attitude, insecurity, discomfort, and optimism,
Malaysia	[5]	210/PLS SEM	Attitude, PBC, and moral norms
China	[57]	253/SEM/AMOS	Attitude, PBC, subjective norms, and residual effect
South-Korea	[9]	1050/SEM/AMOS	Trust, social responsibility, environmental knowledge, and perceived cost.

Source: Authors compilation. Note: PLS-SEM = Partial Least Square Structural Equation Modeling, MGA = Multi-group Analysis, SPSS = Statistical Packages for Social Science, AMOS = Analysis of a Moment Structures.

2.1.4. Environmental Concern

Environmental concern, one of the vital cognitive constructs in the prediction of green products [71], is described by Crosby et al. as "a strong environmental protection attitude" [72]. According to Yadav & Pathak [55], a favorable attitude toward such green things has a beneficial effect on the environment. In an empirical examination of Indian customers, Jaiswal and Kant [71] discovered that environmental concerns were positively connected with attitudes towards green products [71]. In addition, some earlier studies also included this cognitive element in the TPB framework. Chen and Tung merged environmental concern (EC) into an extended TPB framework and discovered that a better EC had a better impact on the attitude, subjective (SN), and PBC of green hotel visitors [73]. Similar results were also obtained by the study conducted by Paul et al. [74], which concluded that EC was favorably linked to the green items and TPB variables. In view of the above discourse, this study coupled the EC with TPB and subsequently advanced the following assumptions:

Hypothesis 5. *Environmental concern has a significant and positive relation to subjective norms.*

Hypothesis 6. *Environmental concern has a significant and positive relation to attitudes toward EEHA.*

Hypothesis 7. *Environmental concern has a significant and positive relation to perceived behavioral control.*

2.1.5. Environmental Knowledge

Consumer knowledge is an important aspect of consumer behavior [75], and its comprehension can, in particular, aid companies in an understanding of the information search and data processing behavior of customers. Environmental knowledge (EK) refers to an individual's environmental expertise, knowledge, and other related problems [76]. Persons with specific ecological knowledge tend to have a positive attitude to ecological behavior and are strongly prepared to take action. Flamm [77] discovered that households having a greater ecological awareness are more likely to purchase energy-efficient vehicles and that research has revealed that ecological knowledge has a favorable impact on customer attitude with respect to green products [47,78,79]. Latif et al. [80] demonstrated that EK influences the attitude of residents towards green products, which in turn affects their intention to purchase green items [6,81]. The following hypotheses are hereby presented:

Hypothesis 8. *Environmental knowledge is significantly and positively related to the attitude of residents towards the purchase of EEHA.*

Hypothesis 9. *Environmental knowledge is significantly and positively related to the intention of residents to buy EEHA.*

2.1.6. Eco-Labeling

Eco-labels are regarded as crucial green marketing tools. As eco-labels are informative instruments for the use, disposal, consumption, and manufacture of items, marketers can seamlessly transmit the ecological benefits of products via eco-labeling [82]. The multidimensional side of eco-labels employed by Prieto-Sandoval et al. [83] encompasses the empirical, geographical, and sectorial views. Since consumers are now very interested in their environment, the possibilities of purchasing eco-labeled products are relatively high. In addition, customer preferences tend to increase when information about energy usage is adequately communicated via well-designed energy labels [68,84,85]. Cho [86] emphasizes the significant impact of ecological and sustainability labeling on a company's attitude. Simon [87] also suggested that environmentally friendly items adhere to the attitude of emerging environmental values. The following theory is subsequently proposed:

Hypothesis 10. *Subjective norms are significantly and positively related to the attitude of residents toward the purchase of EEHA.*

2.1.7. Moderation of Ethical Self-Identity and Environmental Knowledge

Self-identity is the collection of roles portrayed by a person, resulting in a consistent action for self-concept [88]. It is a label used by an individual to bear or identify a particular behavior [89]. Thus, environmental self-identity is described as "an individual's opinion of themselves as an environmentally friendly person" [90]. However, it is important to distinguish between environmental self-identification and environmental identity [91].

Recent research has recognized the mediating effect of consumers' self-identity on the relationship between customer motivation/value and organic food purchase intention [92,93]. To this effect, Van der Werff, Steg, and Keizer [90] revealed that the link between the ecological values, pro-environmental intentions, and behavior of consumers influences green identity. In other words, Environmental knowledge moderates the impact of attitude on the procurement decision of organic foods [94], as a higher level of consumer knowledge results in more comfortable behavior [95]. This, in turn, reduces the impact on the behavior of consumers towards external pressures. In line with the preceding discussion, the following proposition is therefore suggested:

Hypothesis 11. *Green self-identity positively moderates the relationship between the attitude and purchase intention of EEHA.*

Hypothesis 12. *Environmental knowledge positively moderates the relationship between the attitude and purchase intention of EEHA.*

3. Materials and Methods

This is exploratory research based on empirical data (Figure 2). The study utilized original data from a survey of customers to ascertain the factors impacting their green purchasing decisions. It employed a cross-sectional survey design, implying that data was gathered to evaluate the inference of the population at a particular point in time.

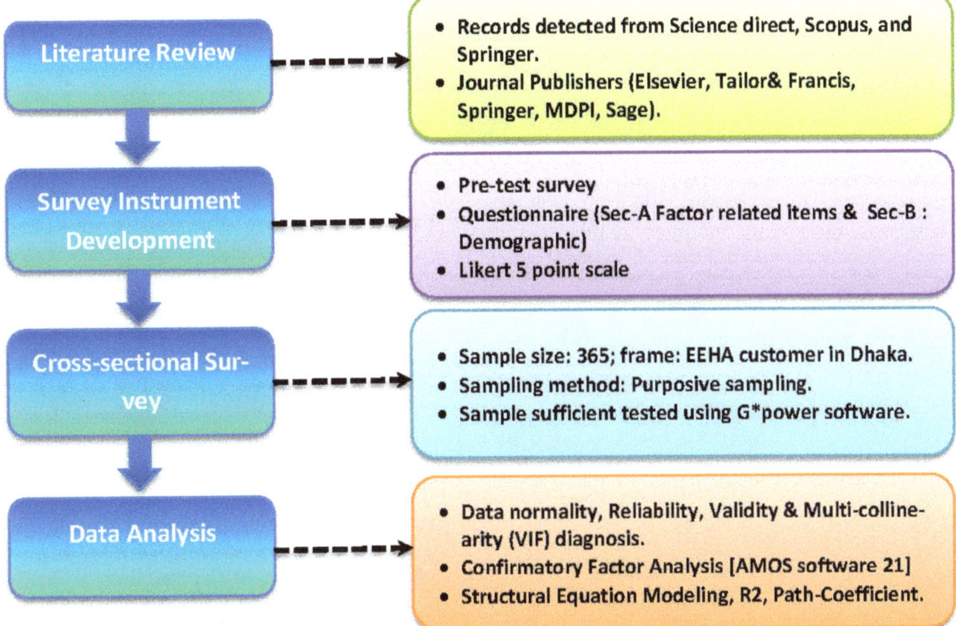

Figure 2. Flowchart of Research Methodology [Source: authors' illustration].

3.1. Sample and Data Collection Procedures

The demography of this study comprises Dhaka customers who are believed to be more informed and cautious in their purchase decisions. Also, compared to other cities in Bangladesh, Dhaka is better positioned in terms of purchasing power. The survey obtained the most data from superstores such as Alesha Mart, Mostafa Mart, and Unimart, and also from Dhaka New Market, which specializes in the sales of home appliances such as electrical kitchen appliances, cooling, and heating appliances, etc. The questionnaire survey was structured.

The study selected respondents based on a purposive sampling strategy. As part of the pre-test study, ten questionnaires were distributed to researchers, and before the final data gathering, a few modifications were implemented based on recommendations from the pre-test study. The minimal sample size was determined using the G*power tool for priory sample size sufficiency [96] and was subsequently used to determine the required number of respondents for the study. Cohen [97] suggested a sample size of 153 for seven independent constructs or predictors (f^2 = 0.15 for effect size, =0.05 for error type 1, and ß = 0.20 for error type 2), while Barclay et al. [98] established a tenfold sample rule in which the maximum number of indicators utilized in the SEM method was multiplied by ten. Going with these criteria, the survey required 240 (10 × 24) responders. However, 400 respondents were reached using the non-probability convenience selection approach to mitigate the potential difficulties associated with small sample sizes. Despite this, 365 samples were subsequently picked after the incomplete inquiries were screened, blotting out incomplete data (23 responses) and screened-out responses (12 responses) in the process. We have considered convenience sampling as a viable alternative owing to its lower costs and convenience of obtaining requisite responses.

3.2. Development of the Questionnaire

Data was obtained via a structured questionnaire, which was adapted from various past research works and contained 29 questions, including a demographic profile in Section 2. Eco-labeling was measured using the items from the conducted study by Neves & Oliveira [45], with three items for assessing the purchase intention being adapted from the conducted study by Ali et al. [99]. The 3-items of environmental concern, environmental knowledge, and attitude were adapted from Li et al. [25]. The green self-identity was culled from Neves & Oliveira [45] & Zheng et al. [45]. The PBC and subjective norms of the constructs were assessed using the items obtained from Zhao et al. [68] and Bhutto et al. [66] accordingly (Appendix A). Furthermore, changes were made to the wordings to fit the organic food perspective. All the items were then measured using a five-point Likert's scale, where (1) represented 'strongly disagree' and (5), 'strongly agree.' In addition, respondents were quizzed on demographic parameters, such as age, profession, education, income, and gender.

3.3. Statistical Analysis

The theoretical framework was investigated using empirical data that was processed with the aid of the SPSS 25, MS-Excel, and AMOS statistical software version 21. There were four indicators analyzed to ensure the validity of the questionnaire survey to determine the effectiveness of the questionnaire items: construct validity, convergent validity, discriminant validity, and questionnaire reliability. Factor loading and cross-loading are both indicators of construct validity. A study's convergent validity can be seen in the factor loading, combined reliability, and average variance extracted. The differences between the average variance extracted and the correlation coefficients between variables reveal discriminant validity. Cronbach's alpha coefficient and composite reliability reflect the questionnaire's reliability. Besides, to assess the data normality, descriptive statistics such as mean, standard deviation, skewness, and kurtosis were used. The study used Variance Inflation Factor (VIF) measures, the severity of multicollinearity, how much the behavior (variance) of an independent variable is influenced, or inflated, by its interaction/correlation with the other independent variables. All parameters were determined by applying the final survey data to SPSS 25 and MS-Excel software.

The structural model was evaluated using the AMOS 21.0 software package, which tested the relationships between the hypothesized constructs. Anderson and Gerbing [100] developed a two-stage structural equation modeling (SEM) technique, in which a confirmatory factor analysis (CFA) was conducted in the first step to verify the reliability and validity of the measurement model. In the second step of the whole structural model, the overall fitness and hypothesized linkages were evaluated using the standardized (β) and p-value regressions using AMOS software relying on the same data set (final survey data). The justification of adopting a two-stage approach, firstly, is to keep the measurement model(s) separate from the structural part during estimation so that they cannot influence each other. Traditionally, maximum likelihood fits all parameters simultaneously. This can lead to interpretation issues for latent variables due to misspecification in the structural model. Secondly, it assists in locating the problems for an unfit model. Convergence issues allow researchers to identify problematic measurement models (s) if it occurs in the first step. If the model fails to converge in the second step, the problem is structural [100].

4. Results

The result section covers the demographic profile of the respondents, reliability, and validity of constructs, followed by the confirmatory factor analysis, structural equation modeling, and moderation analysis.

4.1. Demographics Profile

A high proportion (53.2%) of the responders was males. Similarly, the majority (47.7%) were graduates, while 31.3% had only secondary education or lower. Approximately 44.4% of the responders were private service holders, while 22.6% were government service holders, followed by the self-employed (11.7%) and students (11.1%). With regards to income level, approximately 53.7% earned between $470 and $705, while 20% received between $235 and $470 monthly. The lower-earning was $235 or below constitutes around 4.5% of the total respondents. The age composition of the population revealed a larger proportion (44.6%) of participants as being aged between 30 and 40 years, followed by 40–50 years (25.3%), indicating that the respondents were mature enough to remark on the issues addressed in the study (Table 2).

Table 2. Respondents' Profile.

Aspects	Classification	F	%	Aspects	Classification	F	%
Gender	Male	194	53.2	Income ($1 equals Tk. 85)	Up to $235	16	4.5
	Female	171	46.8		$235–$470	73	20.0
Age	<20	38	10.4		$470-$705	196	53.7
	20–30	55	15.0		$705–$940	57	15.5
	30–40	163	44.6		>940	23	6.3
	40–50	92	25.3	Profession	Student	41	11.1
	>50	17	4.7				
Educational level	No formal education	39	10.6		Entrepreneurs	37	10.2
	Higher Secondary or below	114	31.3		Govt. Job holders	82	22.6
	Graduate	174	47.7		Private Jobs	162	44.4
	Postgraduate and above	38	10.4		Self-employed	43	11.7

$ = indicates the US. Dollar. Source: Selected output of SPSS 25.0.

4.2. Measurement Model (Reliability and Validity)

The assessment of the measurement model was undertaken to estimate the validity and internal consistency of the construct. The construct validity was examined based on the Average Variance Extracted (AVE) and composite reliability. Table 3 shows the reliability and convergent validity measures where item loading is extracted from confirmatory factor analysis, the Cronbach alpha measuring construct reliability calculated using SPSS software 25. The composite reliability and AVE are the validity measure extracted from the MS-excel spreadsheet formula. All the constructs (Table 3) have AVE values higher than 0.5, implying convergent validity [98,101].

Table 4 represents the discriminant validity tested with the popular Fornel Larker method that used correlations assessment via SPSS software. The results of the analysis (Table 4) indicate discriminant validity, as the value of the AVE's square root in the diagonal exceeds other constructs off-diagonal [101]. The value of CR and Cronbach's Alpha–which was greater than 0.7–indicated a good model and was considered highly acceptable for the early stages of the research [102] (Table 3). The constructs of this study are deemed to be statistically satisfactory, as the CR and alpha values exceed the cut-off values earlier stated.

Table 3. Reliability and convergent validity.

Constructs	Item	Item Loading	Cronbach Alpha (α)	Composite Reliability	Average Variance Explained
Eco-Labeling	LB1	0.747			
	LB2	0.777	0.867	0.877	0.707
	LB3	0.979			
Buying Intention	BI1	0.854			
	BI2	0.835	0.868	0.838	0.635
	BI3	0.691			
Perceived Behavioral Control	PBC1	0.893			
	PBC2	0.845	0.907	0.908	0.767
	PBC3	0.889			
Environmental Concern	EC1	0.792			
	EC2	0.905	0.893	0.897	0.745
	EC3	0.888			
Environmental Knowledge	EK1	0.719			
	EK2	0.810	0.796	0.897	0.745
	EK3	0.743			
Subjective Norms	SN1	0.845			
	SN2	0.850	0.864	0.864	0.680
	SN3	0.777			
Attitude	AT1	0.786			
	AT2	0.839	0.907	0.861	0.674
	AT3	0.836			
Green Self–Identity	GSI1:	0.833			
	GSI2:	0.852	0.831	0.876	0.702
	GSI3:	0.828			

Source: Selected output of SPSS 25.0.

Table 4. Discriminant Validity (Fornel Larker Method).

	LB	EK	EC	SN	PBC	AT	GSI	BI
Labeling (LB)	**0.841**							
Environmental Knowledge (EK)	0.584 **	**0.863**						
Environmental Concern (EC)	0.659 **	0.614 **	**0.863**					
Subjective Norms (SN)	0.638 **	0.620 **	0.690 **	**0.825**				
Perceived Behavioral Control (PBC)	0.134 *	0.189 **	0.268 **	0.226 **	**0.876**			
Attitude (AT)	0.676 **	0.675 **	0.715 **	0.729 **	0.206 **	**0.821**		
Green Self–identity (GSI)	0.599 **	0.619 **	0.720 **	0.642 **	0.164 **	0.752 **	**0.838**	
Buying Intention (BI)	0.670 **	0.692 **	0.718 **	0.675 **	0.418 **	0.743 **	0.671 **	**0.797**

(Note: Bold indicates the square root of AVE. ** Correlation is significant at the 0.01 level (2-tailed). *. Correlation is significant at the 0.05 level (2-tailed). Source: Selected output of SPSS 25.0.

For robustness, this study also measured the HTMT value due to its supremacy over Fornell-Larcker in various situations [103]. Table 5 indicates the measurement of HTMT ratio calculated with the help of individual item's correlation and using HTMT formula in the MS Excel software. The outcome recorded a value lower than 0.85/0.90, implying the absence of a discriminant validity problem [103]. Since the present study satisfies the threshold value specified in Table 5, it can be affirmed that the reliability and validity suggested by these analyses are appropriate.

Table 5. Discriminant Validity using Heterotrait-Monotrait Ratio (HTMT).

	LB	EK	EC	SN	PBC	AT	GSI	BI
LB								
EK	0.703							
EC	0.749	0.728						
SN	0.736	0.749	0.786					
PBC	0.152	0.226	0.300	0.258				
AT	0.762	0.794	0.834	0.824	0.230			
GSI	0.720	0.782	0.823	0.777	0.196	0.817		
BI	0.768	0.830	0.828	0.775	0.472	0.831	0.799	

Source: Selected output of SPSS 25.0 and MS-Excel.

4.3. Testing Normality, Multicollinearity, and Coefficient of Determination

Table 6 highlighted data normality, Multicollinearity, and coefficient of determination of constructs. The applied tools, mean, standard deviation, skewness kurtosis, and VIF, were extracted using the descriptive statistics option of SPSS 25. The R square values were obtained from the AMOS 21 output at the time of structural modeling assessment. In terms of normality, the results were good, as no issues were recorded from the variance obtained from the normality assessment. The skewness and kurtosis values were less than ±3 and ±10 [104], respectively (Table 6). The effective technique involving the assessment of the variance inflation factor (VIF) was utilized as recommended by Kleinbaum et al. [105] to determine the presence of multicollinearity amongst the independent variables. The results revealed that the VIF range varies from 1.00 to 2.582, which is substantially below 10. Therefore, this suggests that multicollinearity is not a concern in this investigation.

The R square analyzes the explanatory capacities of models by identifying the endogenous variables highlighted as determining coefficients. According to Cohen [106], the R^2 value of the endogenous variable is substantial when the value is above 0.26 and up to 0.13 is regarded as moderate. However, a value below 0.13 is considered weak. The R^2 estimates of each endogenous value reported in the research in Table 6 are based on the conditions specified by Falk and Miller [107], which demonstrate that the proposed model falls within a high explanatory power range with an exception of the PBC, which has low predicting power.

Table 6. Variance Inflation Factor (VIF) & R^2 Value.

	Mean	Std. Deviation	Skewness	Kurtosis	VIF				R^2	
					AT	BI	PBC	SN	Values	Strength
LB	3.305	0.7534	−0.259	−0.076	2.087					
EK	3.080	0.752	−0.329	−0.438	1.905	1.967				
EC	3.280	0.904	−0.243	−0.064	2.391		1.00	1.00		
SN	3.426	0.779	−0.456	0.323	2.320	2.303				
PBC	2.587	1.089	0.620	−0.652		1.060			0.09	Low
AT	3.342	0.812	−0.485	0.309		2.582			0.70	High
GSI	3.460	0.744	−0.494	0.781					0.61	High
BI	3.298	0.828	−0.300	−0.375					0.77	High

Source: Selected output of SPSS 25.0 and AMOS 21.0.

4.4. Confirmatory Factor Analysis and Common Method Bias

Table 7 explores the CFA and Structural model's fit indices extracted from the AMOS 21 software to verify how fits these models are. In the measurement model, the confirmation of factors was undertaken using the confirmatory factor analysis (CFA), and the resulting CFA model (Table 5) produced good fit indices: $\chi^2/df = 2.531$, Goodness of Fit Index (GFI) = 0.924, Tucker-Lewis Index (TLI) = 0.933, IFI = 0.922, comparative fit index (CFI) = 0.942, NFI = 0.914, and root mean square error of approximation (RMSEA) = 0.069. The t-values corresponding to all the items were significant at a level lower than 5%.

Based on the specified guidelines by Harman [108], common method bias was checked using the Harman's single-factor analysis approach, which relies on the factor analysis method. The single factor represented 31.3% of the variance in the factors, which was lower than the 50% threshold. This affirmed the absence of the common method bias.

Table 7. Results of the CFA and structural model with standards.

Fit Indices	Measurement Values for CFA	Meas. Values for Structural Model	Standards with Sources	
χ^2/df	2.531	2.854	<3	[109]
IFI	0.922	0.938	>0.900	[110]
NFI	0.914	0.909	>0.900	[110]
CFI	0.942	0.938	>0.900	[111]
GFI	0.924	0.911	>0.900	[110]
AGFI	0.918	0.904	>0.900	[101]
TLI	0.933	0.926	$\geq$0.90	[112]
SRMR	0.055	0.061	<0.080	[110]
RMSEA	0.069	0.073	<0.080	[112,113]

Source: Selected output of AMOS 21.0.

4.5. Structural Modeling

The structural model of this analysis is illustrated in Figure 3. As the calculation was successfully carried out in the CFA test of the measurement model, the validation of the structural model was used to check the goodness of fit indices of the proposed model. The outcome of the SEM reveals (Table 7) that the conceptual framework exhibits an excellent data fit ($\chi^2/df = 2.854$). The realized value of the Root Mean Square Error Approximation (RMSEA) was 0.073, which justifies the cut-off value of less than 0.08 [114]. Other fit indices like the CFI, GFI, IFI, and TLI met the standard of 0.9 and higher [113].

Table 8 highlighted the statistic of path model using SEM via AMOS 21.0 software to find the path relationship among constructs as hypothesized. The results (Table 8) indicated that environmental concern influences the subjective norms ($\beta = 0.783$; t = 13.069), perceived behavioral control ($\beta = 0.304$; t = 5.142), and attitude ($\beta = 0.489$; t = 6.029). Likewise, the AMOS output (Table 6) values indicated that the existent relationship between environmental knowledge ($\beta = 0.284$; t = 6.043), subjective norms ($\beta = 0.313$; t = 3.924), eco-labeling ($\beta = 0.206$; t = 5.057) and attitude were significant. In addition to that, environmental knowledge ($\beta = 0.295$; t = 5.810), subjective norms ($\beta = 0.172$; t = 2.533; $p < 0.05$), attitude ($\beta = 0.497$; t = 6.653), perceived behavioral control ($\beta = 0.324$; t = 7.636) and buying intention were also found to be significant. Therefore, we accept hypotheses 1 to 10, all of which are significance at the 1% level except for H_7, which exhibits significance at the 5% level.

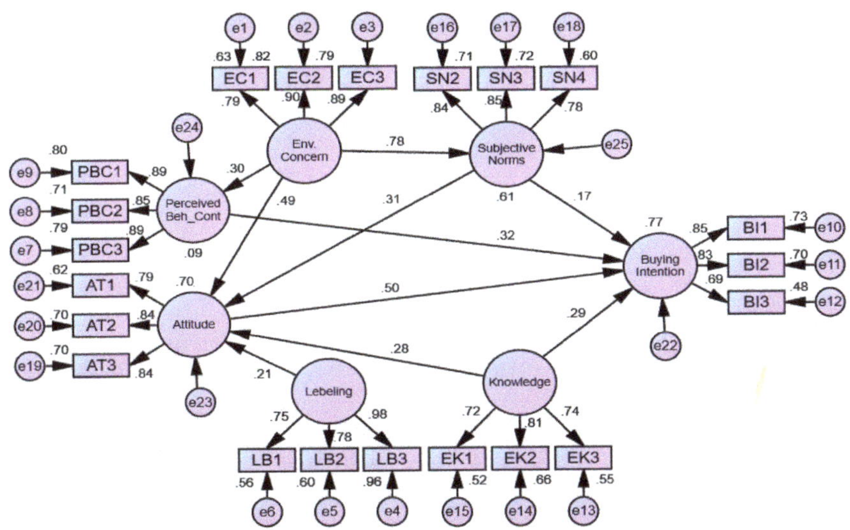

Figure 3. Structural Model [Source: Selected output of AMOS 21.0].

Table 8. Structural Model and the Hypothesis Testing Result.

Hypotheses	STD Beta	STD Error	*t*-Values	*p*-Values	Significance (*p* < 0.05)
H1: EC→ SN	0.783	0.055	13.069 ***	0.000	Supported
H2: EC→ AT	0.489	0.065	6.029 ***	0.000	Supported
H3: EC→ PBC	0.304	0.079	5.142 ***	0.000	Supported
H4: EK→ AT	0.284	0.045	6.043 ***	0.000	Supported
H5: EK→ BI	0.295	0.049	5.810 ***	0.000	Supported
H6: SN→ AT	0.313	0.071	3.924 ***	0.000	Supported
H7: SN→BI	0.172	0.060	2.533 **	0.011	Supported
H8: LB→AT	0.206	0.031	5.057 ***	0.000	Supported
H9: AT→ BI	0.497	0.075	6.653 ***	0.000	Supported
H10: PBC→ BI	0.324	0.026	7.636 ***	0.000	Supported
H11: GSI*AT → BI	0.289	0.039	6.324 ***	0.000	Supported
H12: EK*AT → BI	0.018	0.172	0.606	0.237	Not Supported

** Significant at 5% level, *** Significant at 1% level, Source: Selected output of AMOS 21.0.

4.6. The Moderation of Green Self-Identity and Environmental Knowledge

The moderation effect was assessed based on the interaction effects of the variables. Figure 4 illustrates the moderation relationship among hypothesized variables using the Sobel test approach applying MS-Excel Spreadsheet. The study results (Figure 4 and Table 8) revealed that the green self-identity moderates the association between the attitude and purchase intention (β = 0.289, t = 6.324, p < 0.01), while in contrast, the environmental knowledge does not moderate the association between attitude (β = 0.018, t = 0.606, p > 0.05) and purchase intention. Therefore, H$_{11}$ is validated while H$_{12}$ is rejected. A higher green self-identity influences buying intention when a consumer is more inclined to purchase EEHA.

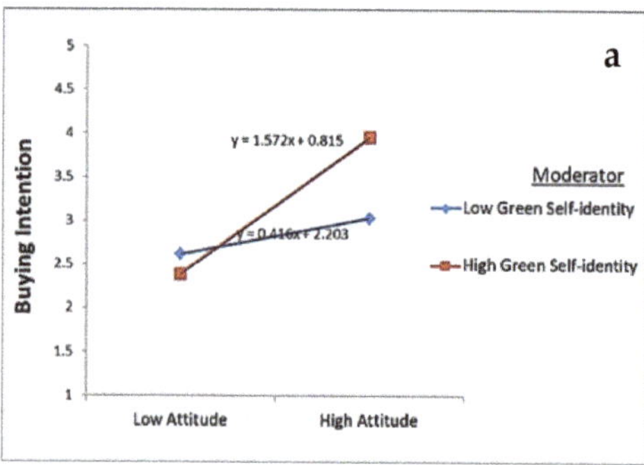

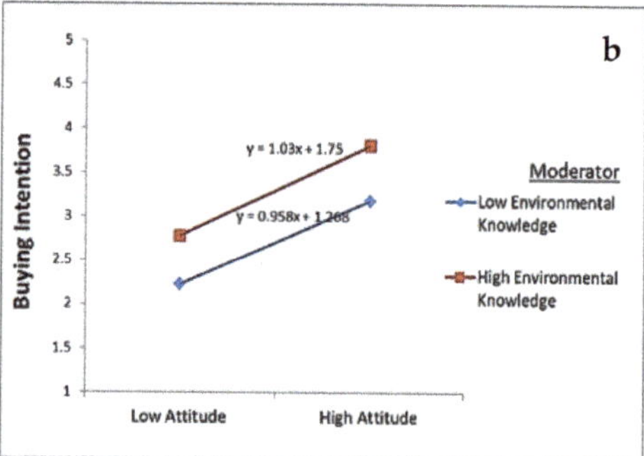

Figure 4. Interaction of (**a**) green self-identity is significant (**b**) environmental knowledge is insignificant between attitude and buying intention relationships, Source: Sobel test output.

5. Discussion

The study endeavored to incorporate additional cognitive constructs such as environmental concern, environmental knowledge, eco-labeling, and green self-identity along with the original TPB variables. In this proposed model, the R^2 value of behavioral intention is 0.77, which is greater than the values of 0.319 found in the original TPB model [115]. These results showed that because extended TPBs can perceive behavioral intention, the proposed model is generally adequate, comprehensive, and functional for comprehending EEHA product purchases.

According to the outcome of this study, environmental concern influences the subjective norms, perceived behavioral control, and attitude towards the purchase of EEHA in Bangladesh, thus validating Hypotheses 1–3. This signifies that the deeper the environmental concern expressed by the customers, the greater the external influence on purchase, internal motivation, and attitude towards the purchase of energy-efficient products, hence corroborating the previous research [23,24,74,116]. The results (H4 and H5) also revealed that environmental knowledge led to a more favorable attitude and intention towards the purchase of energy-efficient products in Bangladesh, which is complementary to the conducted study by China's Liu et al. [117]. This implies that customers tend to become more self-reliant in formulating opinions and making judgments when they are equipped with knowledge about the EEHA in Bangladesh perspectives.

The empirical assessment in this study identified the subjective norms as being of significant influence on customers' attitudes towards the purchase and buying intention of energy-efficient home appliances (H6 and H7). This result is in line with prior studies [118,119], which connotes that a higher social influence will result in a higher buying attitude and purchase intention of EEHA products in Bangladesh. It is noted that subjective norm is observed as the lowest relationship strength (0.172) with behavioral intention. Likewise, consistent with past studies [87], energy labeling is a determinant factor of attitude, which validates Hypothesis 8. This result stresses that the access of people to information about a product's energy efficiency via the use of labeling will result in their tendency to make more purchases. As expected and in line with previous works of literature [23,24,66,99], attitude and perceived behavioral control are key drivers of the purchase intention of EEHA products. This validates Hypotheses 9 and 10, indicating that a favorable attitude and higher PBC enhance the purchase intention of EEHA products. However, attitude is found as the highest strength (0.497) of relationship with the behavioral intention followed by the perceived behavioral control (0.324).

This study endorses the moderating role of the green self-identity between attitude and buying intention, implying that an individual having a greater level of green self-identity holds a stronger conviction to be more eco-friendly and is thus more inclined to act in an eco-friendly manner (H11). However, contrary to the study conducted by Cinjarevi'c et al. [94], environmental knowledge failed to moderate the positive association between the attitude and purchase intention of EEHA products (H12).

6. Implications of the Study

This research has contributed to the extant literature in numerous ways. First, based on the researcher's knowledge, this is the first research to address the buying intention exhibited by Bangladeshis towards energy-efficient home appliance products. Although some research on this issue abounds in South Asian countries and other regions, studies from the perspective of developing countries such as Bangladesh are lacking. Second, the attitude-intention gap is a pressing issue experienced in individual behavior models, particularly eco-friendly products [37]. However, consumers express a highly positive attitude towards eco-product when quizzed, indicating a deficiency in intention-behavior, as many of these attitudes do not transform into intention or behavior. The current study addressed the issue to find possible reasons for such gaps in Bangladesh by incorporating green self-identity and environmental knowledge variables in the research framework.

Third, the current study contributes to academia by establishing the moderating roles of green self-identity between attitude and buying intention. Existing research works mainly regard the green self-identity as having a moderating effect on the consumption value, intention or savings, label, operational and maintenance co-benefits, and intention, while overlooking the moderating impact of the green self-identity between attitude and intention using the TPB model. To the best of the author's knowledge, this study is one of the first to empirically examine how green self-identity moderates attitude-intention relationships. This knowledge will help to provide an understanding of the role of consumers' green emotion to bridge the attitude-intention gaps.

Fourth, this study contributes to recent studies that found subjective norm to be a determinant of both attitude and behavioral intentions [118,119] in a single model in other fields. Rarely, the subjective norm was observed to be simultaneously significant with the attitude and intention, and none of the studies on the purchase of energy-saving home appliance products claimed to have both impacts simultaneously. This study established subjective norms as a very important predictor of the behavioral intention of the purchase of EEHA. This will strengthen the schools of thought that are of the belief that external influence (subjective norms) matters in the implementation of purchase decisions and those that advocate for the enhancement of community engagement by the companies in addition to the presence in the markets.

This study offers numerous practical implications for managers and policymakers. As the eco-label was proven to be associated with purchase behavior, the energy label must be placed on all equipment since it is a vital factor that individuals regard as premium. Therefore, the energy label must be visually appealing and indicate the energy categorization of the equipment. This recommendation may be more geared toward energy agencies. Second, the perceived behavioral control and green self-identity significantly influence the intention of residents to purchase EEHA. The buyers' belief in their ability to make decisions and control and their green emotions should be strategically targeted. Therefore, promotional campaigns should be actively undertaken by the seller to induce prospective buyers. The promotional activities could range from publicity, advertisement, and personal sales, to sales promotion like discounts to enable buyers to remain informed about the product and spur them to make a purchase decision.

Third, according to confirmed Hypotheses 1 to 5, environmental concern and environmental knowledge influence the buying attitude and subjective norm and ultimately influence the buying intention of EEHA. Equipped with the role of a private enterprise, the government should promote environmental education and awareness programs. This is critical to the enlightenment of residents to the essence of energy-saving and cutting emissions, logical use of mass platforms to promote knowledge about EEHA and ecological protection, and a continuous effort to change the attitude of residents toward EEHA and increase their purchase intention. Besides, with the proper regulations and policies on the development of local manufacturing companies of EEHA, a subsidiary policy should be undertaken to promote and enable startups to endure competition. It may involve the rendering of technical help, monetary incentives, and exemptions from taxes on energy-efficient imported technologies and equipment.

7. Conclusions and Limitations

This study investigated the factors influencing the buying intention of EEHA in Bangladesh. It confirmed that the environmental knowledge, subjective norms, attitude, and perceived behavioral control significantly affected consumers' buying intention of energy-efficient home appliances. The result revealed a significant relationship among environmental concern, environmental knowledge, subjective norms, eco-labeling, and attitude towards purchases. It also discovered that the green self-identity moderates the relationship between the buying intention and attitude of EEHA products, while environmental knowledge failed to moderate the same relationship.

This study examined the purchase intentions of green products and may be further extended to include the repurchase intention in the future to provide an insight into how consumers might be sustained. In this study, we used behavioral intentions to consume EEHA products. Future research can incorporate actual behavior to provide an understanding of how much intention transforms into actual activity. Finally, this research was limited to a sample of energy-efficient product customers in Dhaka, Bangladesh (Urban population). Therefore, to acquire a better understanding of the breadth and depth of consumer intention towards the purchase of EEHA in Bangladesh, a sample from other cities, including smaller towns and rural areas, is essential.

Author Contributions: Conceptualization, A.B.S., Y.L. and M.M.; methodology, A.B.S., M.M. and Y.L.; software, A.B.S. and M.M.; validation, Y.L., A.B.S. and M.M.; investigation, M.M., A.B.S. and Y.L.; resources, A.B.S., Y.L. and M.M.; data curation, M.M., A.B.S. and Y.L.; writing—original draft preparation, Y.L., A.B.S. and M.M.; writing—review and editing, Y.L., X.W. and M.M.; visualization, A.B.S., X.W. and Y.L.; supervision, X.W. and M.M.; Funding acquisition, Y.L. All authors have read and agreed to the published version of the manuscript.

Funding: This study was funded by the National Social Science Fund of China (grant no. 19BJY177).

Institutional Review Board Statement: Ethical review and approval were waived for this study due to the fact that there is no institutional review board or committee in Bangladesh. Besides, the study was conducted as per the guidelines of the Declaration of Helsinki. The research questionnaire was anonymous, and no personal information was gathered.

Informed Consent Statement: Oral consent was obtained from all individuals involved in this study.

Data Availability Statement: The data that support the findings of this study are available from the corresponding authors (A.B.S.) upon reasonable request.

Conflicts of Interest: The authors declare no conflict of interest.

Appendix A

Table A1. Survey Instruments.

Constructs	Source
Eco-Labeling	[45]
LB1. The energy label is critical when purchasing a household appliance.	
LB2. When I purchase a household appliance, I carefully read the energy label.	
LB3. I am more receptive to purchasing a household appliance with a high-energy efficiency rating (above C, i.e., A or B)	

Table A1. *Cont.*

Constructs	Source
Buying Intention	[99]
BI1 I plan to get one that is energy efficient when I need to purchase home appliances.	
BI2 I intend to make a concerted effort to acquire energy-efficient appliances.	
BI3 I will switch from conventional to energy-efficient appliances.	
Perceived Behavioral Control	[68]
PBC1 Purchasing energy-efficient equipment is simple for me.	
PBC2 I believe I am financially capable of purchasing energy-efficient appliances.	
PBC3 Whether or not I choose and acquire energy-efficient appliances in my daily life is entirely up to me.	
Environmental Concern	[24]
EC1: The natural balance is precarious and susceptible to disruption.	
EC2: When humans disturb nature, disastrous results often occur.	
EC3: To thrive, humans must live in harmony with nature.	
Environmental Knowledge	[24]
EK1: I can determine whether the appliances I purchased are environmentally friendly.	
EK2: I am more knowledgeable about recycling than the average person.	
EK3: I am well knowledgeable about environmental issues.	
Subjective Norms	[66]
SN1: The majority of people who matter to me believe that I should invest in EEHA.	
SN2: Using energy-efficient appliances is a social trend.	
SN3: People whose opinion I respect would buy energy-efficient appliances instead of conventional ones.	
Attitude	[24]
ATT1: I believe that purchasing EEHA is an excellent habit.	
ATT2: I believe that purchasing EEHA is a wise investment.	
AT3: I believe that shopping for EEHA is a satisfying experience.	
Green Self–Identity	[45,120]
GSI1: Environmental protection starts with me	
GSI2: I consider myself a "green consumer."	
GSI3: Environmental protection is the government's responsibility, not mine.	

References

1. Ali, S.; Ullah, H.; Akbar, M.; Akhtar, W.; Zahid, H. Determinants of consumer intentions to purchase energy-saving household products in Pakistan. *Sustainability* **2019**, *11*, 1462. [CrossRef]
2. Sorrell, S. Reducing energy demand: A review of issues, challenges and approaches. *Renew. Sustain. Energy Rev.* **2015**, *47*, 74–82. [CrossRef]
3. Zhou, K.; Yang, S. Understanding household energy consumption behavior: The contribution of energy big data analytics. *Renew. Sustain. Energy Rev.* **2016**, *56*, 810–819. [CrossRef]
4. Ek, K.; Söderholm, P. The devil is in the details: Household electricity saving behavior and the role of information. *Energy Policy* **2010**, *38*, 1578–1587. [CrossRef]
5. Tan, C.-S.; Ooi, H.-Y.; Goh, Y.-N. A moral extension of the theory of planned behavior to predict consumers' purchase intention for energy-efficient household appliances in Malaysia. *Energy Policy* **2017**, *107*, 459–471. [CrossRef]
6. Wang, Z.; Zhang, B.; Li, G. Determinants of energy-saving behavioral intention among residents in Beijing: Extending the theory of planned behavior. *J. Renew. Sustain. Energy* **2014**, *6*, 53127. [CrossRef]
7. Mills, B.; Schleich, J. Analysis of Existing Data: Determinants for the Adoption of Energy-Efficient Household Appliances in Germany. In *Sustainable Energy Consumption in Residential Buildings*; Springer: Berlin/Heidelberg, Germany, 2013; pp. 39–67.

8.   Frederiks, E.R.; Stenner, K.; Hobman, E.V.; Fischle, M. Evaluating energy behavior change programs using randomized controlled trials: Best practice guidelines for policymakers. *Energy Res. Soc. Sci.* **2016**, *22*, 147–164. [CrossRef]

9.   Park, E.; Kwon, S.J. What motivations drive sustainable energy-saving behavior?: An examination in South Korea. *Renew. Sustain. Energy Rev.* **2017**, *79*, 494–502. [CrossRef]

10. Ahmed, A.S.; Kumari, D.R.; Neeraja, T. Purchase Behavior of Women Consumers, towards Energy Efficient Household Products. Doctoral Dissertation, Univeristy Rajendranagar, Hyderabad, India, 2017.

11. Joshi, G.; Sen, V.; Kunte, M. Do Star Ratings Matter?: A Qualitative Study on Consumer Awareness and Inclination to Purchase Energy-Efficient Home Appliances. *Int. J. Soc. Ecol. Sustain. Dev.* **2020**, *11*, 40–55. [CrossRef]

12. Poortinga, W.; Steg, L.; Vlek, C.; Wiersma, G. Household preferences for energy-saving measures: A conjoint analysis. *J. Econ. Psychol.* **2003**, *24*, 49–64. [CrossRef]

13. Wai, C.W.; Mohammed, A.H.; Alias, B. Energy conservation: A conceptual framework of energy awareness development process. *Malaysian J. Real Estate* **2009**, *1*, 58–67.

14. Oikonomou, V.; Becchis, F.; Steg, L.; Russolillo, D. Energy saving and energy efficiency concepts for policy making. *Energy Policy* **2009**, *37*, 4787–4796. [CrossRef]

15. Abrahamse, W.; Steg, L. How do socio-demographic and psychological factors relate to households' direct and indirect energy use and savings? *J. Econ. Psychol.* **2009**, *30*, 711–720. [CrossRef]

16. Niemeyer, S. Consumer voices: Adoption of residential energy-efficient practices. *Int. J. Consum. Stud.* **2010**, *34*, 140–145. [CrossRef]

17. Pothitou, M.; Hanna, R.F.; Chalvatzis, K.J. Environmental knowledge, pro-environmental behaviour and energy savings in households: An empirical study. *Appl. Energy* **2016**, *184*, 1217–1229. [CrossRef]

18. Tanner, C.; Kast, S.W. Promoting Sustainable Consumption: Determinants of Green Purchases by Swiss Consumers. *Psychol. Mark.* **2003**, *20*, 883–902. [CrossRef]

19. Baldini, M.; Trivella, A.; Wente, J.W. The impact of socioeconomic and behavioural factors for purchasing energy-efficient household appliances: A case study for Denmark. *Energy Policy* **2018**, *120*, 503–513. [CrossRef]

20. Jain, S.K.; Kaur, G. Role of socio-demographics in segmenting and profiling green consumers: An exploratory study of consumers in India. *J. Int. Consum. Mark.* **2006**, *18*, 107–146. [CrossRef]

21. Jeong, G.; Kim, Y. The effects of energy efficiency and environmental labels on appliance choice in South Korea. *Energy Effic.* **2015**, *8*, 559–576. [CrossRef]

22. Lin, J.; Lobo, A.; Leckie, C. The role of benefits and transparency in shaping consumers' green perceived value, self-brand connection and brand loyalty. *J. Retail. Consum. Serv.* **2017**, *35*, 133–141. [CrossRef]

23. Zhang, L.; Fan, Y.; Zhang, W.; Zhang, S. Extending the theory of planned behavior to explain the effects of cognitive factors across different kinds of green products. *Sustainability* **2019**, *11*, 4222. [CrossRef]

24. Li, G.; Li, W.; Jin, Z.; Wang, Z. Influence of environmental concern and knowledge on households' willingness to purchase energy-efficient appliances: A case study in Shanxi, China. *Sustainability* **2019**, *11*, 1073. [CrossRef]

25. Hua, L.; Wang, S. Antecedents of consumers' intention to purchase energy-efficient appliances: An empirical study based on the technology acceptance model and theory of planned behavior. *Sustainability* **2019**, *11*, 2994. [CrossRef]

26. Chen, M.-F. Extending the theory of planned behavior model to explain people's energy savings and carbon reduction behavioral intentions to mitigate climate change in Taiwan—moral obligation matters. *J. Clean. Prod.* **2016**, *112*, 1746–1753. [CrossRef]

27. Yuriev, A.; Dahmen, M.; Paillé, P.; Boiral, O.; Guillaumie, L. Pro-environmental behaviors through the lens of the theory of planned behavior: A scoping review. *Resour. Conserv. Recycl.* **2020**, *155*, 104660. [CrossRef]

28. Davies, J.; Foxall, G.R.; Pallister, J. Beyond the intention–behaviour mythology: An integrated model of recycling. *Mark. Theory* **2002**, *2*, 29–113. [CrossRef]

29. Ertz, M.; Huang, R.; Jo, M.-S.; Karakas, F.; Sarigöllü, E. From single-use to multi-use: Study of consumers' behavior toward consumption of reusable containers. *J. Environ. Manag.* **2017**, *193*, 334–344. [CrossRef]

30. Ajzen, I. *Constructing a TPB Questionnaire: Conceptual and Methodological Considerations*; CiteSeer: Princeton, NJ, USA, 2002.

31. Rhodes, R.E.; Courneya, K.S. Investigating multiple components of attitude, subjective norm, and perceived control: An examination of the theory of planned behaviour in the exercise domain. *Br. J. Soc. Psychol.* **2003**, *42*, 129–146. [CrossRef]

32. Ajzen, I. The theory of planned behavior. *Organ. Behav. Hum. Decis. Process.* **1991**, *50*, 179–211. [CrossRef]

33. Ajzen, I. The theory of planned behavior: Frequently asked questions. *Hum. Behav. Emerg. Technol.* **2020**, *2*, 314–324. [CrossRef]

34. Al Mamun, A.; Mohamad, M.R.; Bin Yaacob, M.R.; Mohiuddin, M. Intention and behavior towards green consumption among low-income households. *J. Environ. Manag.* **2018**, *227*, 73–86. [CrossRef]

35. Sreen, N.; Purbey, S.; Sadarangani, P. Impact of culture, behavior and gender on green purchase intention. *J. Retail. Consum. Serv.* **2018**, *41*, 177–189. [CrossRef]

36. Kaffashi, S.; Shamsudin, M.N. Transforming to a low carbon society; an extended theory of planned behaviour of Malaysian citizens. *J. Clean. Prod.* **2019**, *235*, 1255–1264. [CrossRef]

37. Wiederhold, M.; Martinez, L.F. Ethical consumer behaviour in Germany: The attitude-behaviour gap in the green apparel industry. *Int. J. Consum. Stud.* **2018**, *42*, 419–429. [CrossRef]

38. Gleim, M.; Lawson, S.J. Spanning the gap: An examination of the factors leading to the green gap. *J. Consum. Mark.* **2014**, *31*, 503–514. [CrossRef]

39. Vermeir, I.; Verbeke, W. Impact of values, involvement and perceptions on consumer attitudes and intentions towards sustainable consumption. *J. Agric. Environ. Ethics* **2006**, *19*, 169–194. [CrossRef]

40. Van der Werff, E.; Steg, L.; Keizer, K. It is a moral issue: The relationship between environmental self-identity, obligation-based intrinsic motivation and pro-environmental behaviour. *Glob. Environ. Chang.* **2013**, *23*, 1258–1265. [CrossRef]

41. Nigbur, D.; Lyons, E.; Uzzell, D. Attitudes, norms, identity and environmental behaviour: Using an expanded theory of planned behaviour to predict participation in a kerbside recycling programme. *Br. J. Soc. Psychol.* **2010**, *49*, 259–284. [CrossRef]

42. Gatersleben, B.; Murtagh, N.; Abrahamse, W. Values, identity and pro-environmental behaviour. *Contemp. Soc. Sci.* **2014**, *9*, 374–392. [CrossRef]

43. Kashima, Y.; Paladino, A.; Margetts, E.A. Environmentalist identity and environmental striving. *J. Environ. Psychol.* **2014**, *38*, 64–75. [CrossRef]

44. Qasim, H.; Yan, L.; Guo, R.; Saeed, A.; Ashraf, B.N. The defining role of environmental self-identity among consumption values and behavioral intention to consume organic food. *Int. J. Environ. Res. Public Health* **2019**, *16*, 1106. [CrossRef] [PubMed]

45. Neves, J.; Oliveira, T. Understanding energy-efficient heating appliance behavior change: The moderating impact of the green self-identity. *Energy* **2021**, *225*, 120169. [CrossRef]

46. Mostafa, M.M. Gender differences in Egyptian consumers' green purchase behaviour: The effects of environmental knowledge, concern and attitude. *Int. J. Consum. Stud.* **2007**, *31*, 220–229. [CrossRef]

47. Polonsky, M.J.; Vocino, A.; Grau, S.L.; Garma, R.; Ferdous, A.S. The impact of general and carbon-related environmental knowledge on attitudes and behaviour of US consumers. *J. Mark. Manag.* **2012**, *28*, 238–263. [CrossRef]

48. Laroche, M.; Tomiuk, M.; Bergeron, J.; Barbaro-Forleo, G. Cultural differences in environmental knowledge, attitudes, and behaviours of Canadian consumers. *Can. J. Adm. Sci. Can. Sci. l'Adm.* **2002**, *19*, 267–282. [CrossRef]

49. Kollmuss, A.; Agyeman, J. Mind the gap: Why do people act environmentally and what are the barriers to pro-environmental behavior? *Environ. Educ. Res.* **2002**, *8*, 239–260. [CrossRef]

50. Kim, E.; Ham, S.; Yang, I.S.; Choi, J.G. The roles of attitude, subjective norm, and perceived behavioral control in the formation of consumers' behavioral intentions to read menu labels in the restaurant industry. *Int. J. Hosp. Manag.* **2013**, *35*, 203–213. [CrossRef]

51. Tarkiainen, A.; Sundqvist, S. Subjective norms, attitudes and intentions of Finnish consumers in buying organic food. *Br. Food J.* **2005**, *107*, 808–822. [CrossRef]

52. Ajzen, I.; Fishbein, M. *Understing Attitudes and Predicting Social Behaviour*; Prentice-Hall: Englewood Cliffs, NJ, USA, 1980.

53. Beck, L.; Ajzen, I. Predicting dishonest actions using the theory of planned behavior. *J. Res. Personal.* **1991**, *25*, 285–301. [CrossRef]

54. Verma, V.K.; Chandra, B. An application of theory of planned behavior to predict young Indian consumers' green hotel visit intention. *J. Clean. Prod.* **2018**, *172*, 1152–1162. [CrossRef]

55. Yadav, R.; Pathak, G.S. Determinants of consumers' green purchase behavior in a developing nation: Applying and extending the theory of planned behavior. *Ecol. Econ.* **2017**, *134*, 114–122. [CrossRef]

56. Nguyen, T.N.; Lobo, A.; Greenland, S. Energy efficient household appliances in emerging markets: The influence of consumers' values and knowledge on their attitudes and purchase behaviour. *Int. J. Consum. Stud.* **2017**, *41*, 167–177. [CrossRef]

57. Wang, Z.; Wang, X.; Guo, D. Policy implications of the purchasing intentions towards energy-efficient appliances among China's urban residents: Do subsidies work? *Energy Policy* **2017**, *102*, 430–439. [CrossRef]

58. Ha, H.; Janda, S. Predicting consumer intentions to purchase energy-efficient products. *J. Consum. Mark.* **2012**, *29*, 461–469. [CrossRef]

59. Hori, S.; Kondo, K.; Nogata, D.; Ben, H. The determinants of household energy-saving behavior: Survey and comparison in five major Asian cities. *Energy Policy* **2013**, *52*, 354–362. [CrossRef]

60. Wang, Z.; Sun, Q.; Wang, B.; Zhang, B. Purchasing intentions of Chinese consumers on energy-efficient appliances: Is the energy efficiency label effective? *J. Clean. Prod.* **2019**, *238*, 117896. [CrossRef]

61. Zainudin, N.; Siwar, C.; Choy, E.A.; Chamhuri, N. Evaluating the Role of Energy Efficiency Label on Consumers' Purchasing Behaviour. *APCBEE Procedia* **2014**, *10*, 326–330. [CrossRef]

62. Wang, P.; Liu, Q.; Qi, Y. Factors influencing sustainable consumption behaviors: A survey of the rural residents in China. *J. Clean. Prod.* **2014**, *63*, 152–165. [CrossRef]

63. Alam, S.S.; Hashim, N.H.N.; Rashid, M.; Omar, N.A.; Ahsan, N.; Ismail, M.D. Small-scale households renewable energy usage intention: Theoretical development and empirical settings. *Renew. Energy* **2014**, *68*, 255–263. [CrossRef]

64. Gaspar, R.; Antunes, D. Energy efficiency and appliance purchases in Europe: Consumer profiles and choice determinants. *Energy Policy* **2011**, *39*, 7335–7346. [CrossRef]

65. Wang, Z.; Zhang, B.; Yin, J.; Zhang, Y. Determinants and policy implications for household electricity-saving behaviour: Evidence from Beijing, China. *Energy Policy* **2011**, *39*, 3550–3557. [CrossRef]

66. Bhutto, M.Y.; Liu, X.; Soomro, Y.A.; Ertz, M.; Baeshen, Y. Adoption of energy-efficient home appliances: Extending the theory of planned behavior. *Sustainability* **2021**, *13*, 250. [CrossRef]

67. Zou, B.; Mishra, A.K. Appliance usage and choice of energy-efficient appliances: Evidence from rural Chinese households. *Energy Policy* **2020**, *146*, 111800. [CrossRef]

68. Zhao, C.; Zhang, M.; Wang, W. Exploring the influence of severe haze pollution on residents' intention to purchase energy-saving appliances. *J. Clean. Prod.* **2019**, *212*, 1536–1543. [CrossRef]

69. Joshi, G.Y.; Sheorey, P.A.; Gandhi, A.V. Analyzing the barriers to purchase intentions of energy efficient appliances from consumer perspective. *Benchmarking Int. J.* **2019**, *26*, 1565–1580. [CrossRef]

70. Kim, W.; Ko, S.; Oh, M.; Choi, I.; Shin, J. Is an incentive policy for energy efficient products effective for air purifiers? The case of South Korea. *Energies* **2019**, *12*, 1664. [CrossRef]

71. Jaiswal, D.; Kant, R. Green purchasing behaviour: A conceptual framework and empirical investigation of Indian consumers. *J. Retail. Consum. Serv.* **2018**, *41*, 60–69. [CrossRef]

72. Crosby, L.A.; Gill, J.D.; Taylor, J.R. Consumer/voter behavior in the passage of the Michigan container law. *J. Mark.* **1981**, *45*, 19–32. [CrossRef]

73. Chen, M.F.; Tung, P.J. Developing an extended Theory of Planned Behavior model to predict consumers' intention to visit green hotels. *Int. J. Hosp. Manag.* **2014**, *36*, 221–230. [CrossRef]

74. Paul, J.; Modi, A.; Patel, J. Predicting green product consumption using theory of planned behavior and reasoned action. *J. Retail. Consum. Serv.* **2016**, *29*, 123–134. [CrossRef]

75. Oh, K.; Abraham, L. Effect of knowledge on decision making in the context of organic cotton clothing. *Int. J. Consum. Stud.* **2016**, *40*, 66–74. [CrossRef]

76. Chan, R.Y.K.; Lau, L.B.Y. Antecedents of green purchases: A survey in China. *J. Consum. Mark.* **2000**, *17*, 338–357. [CrossRef]

77. Flamm, B. The impacts of environmental knowledge and attitudes on vehicle ownership and use. *Transp. Res. Part D Transp. Environ.* **2009**, *14*, 272–279. [CrossRef]

78. Mostafa, M.M. Shades of green: A psychographic segmentation of the green consumer in Kuwait using self-organizing maps. *Expert Syst. Appl.* **2009**, *36*, 11030–11038. [CrossRef]

79. Barber, N.; Taylor, C.; Strick, S. Wine consumers' environmental knowledge and attitudes: Influence on willingness to purchase. *Int. J. Wine Res.* **2009**, *1*, 59–72. [CrossRef]

80. Latif, S.A.; Omar, M.S.; Bidin, Y.H.; Awang, Z. Role of environmental knowledge in creating pro-environmental residents. *Procedia-Soc. Behav. Sci.* **2013**, *105*, 866–874. [CrossRef]

81. Vicente-Molina, M.A.; Fernández-Sáinz, A.; Izagirre-Olaizola, J. Environmental knowledge and other variables affecting pro-environmental behaviour: Comparison of university students from emerging and advanced countries. *J. Clean. Prod.* **2013**, *61*, 130–138. [CrossRef]

82. Atkinson, L.; Rosenthal, S. Signaling the green sell: The influence of eco-label source, argument specificity, and product involvement on consumer trust. *J. Advert.* **2014**, *43*, 33–45. [CrossRef]

83. Prieto-Sandoval, V.; Alfaro, J.A.; Mejía-Villa, A.; Ormazabal, M. ECO-labels as a multidimensional research topic: Trends and opportunities. *J. Clean. Prod.* **2016**, *135*, 806–818. [CrossRef]

84. Shen, J.; Saijo, T. Does an energy efficiency label alter consumers' purchasing decisions? A latent class approach based on a stated choice experiment in Shanghai. *J. Environ. Manag.* **2009**, *90*, 3561–3573. [CrossRef]

85. Stadelmann, M.; Schubert, R. How do different designs of energy labels influence purchases of household appliances? A field study in Switzerland. *Ecol. Econ.* **2018**, *144*, 112–123. [CrossRef]

86. Cho, E.; Gupta, S.; Kim, Y. Style consumption: Its drivers and role in sustainable apparel consumption. *Int. J. Consum. Stud.* **2015**, *39*, 661–669. [CrossRef]

87. Simon, F.L. Marketing green products in the triad. *Columbia J. World Bus.* **1992**, *27*, 268–285.

88. Stets, J.E.; Burke, P.J. Identity theory and social identity theory. *Soc. Psychol. Q.* **2000**, *63*, 224–237. [CrossRef]

89. Conner, M.; Armitage, C.J. Extending the theory of planned behavior: A review and avenues for further research. *J. Appl. Soc. Psychol.* **1998**, *28*, 1429–1464. [CrossRef]

90. Van der Werff, E.; Steg, L.; Keizer, K. The value of environmental self-identity: The relationship between biospheric values, environmental self-identity and environmental preferences, intentions and behaviour. *J. Environ. Psychol.* **2013**, *34*, 55–63. [CrossRef]

91. Stets, J.E.; Biga, C.F. Bringing identity theory into environmental sociology. *Sociol. Theory* **2003**, *21*, 398–423. [CrossRef]

92. Cook, A.J.; Kerr, G.N.; Moore, K. Attitudes and intentions towards purchasing GM food. *J. Econ. Psychol.* **2002**, *23*, 557–572. [CrossRef]

93. Sparks, P.; Shepherd, R. Self-Identity and the Theory of Planned Behavior: Assesing the Role of Identification with "Green Consumerism". *Soc. Psychol. Q.* **1992**, *55*, 388–399. [CrossRef]

94. Činjarević, M.; Agić, E.; Peštek, A. When consumers are in doubt, you better watch out! the moderating role of consumer skepticism and subjective knowledge in the context of organic food consumption. *Zagreb Int. Rev. Econ. Bus.* **2018**, *21*, 1–14. [CrossRef]

95. Raju, P.S.; Lonial, S.C.; Mangold, W.G. Differential effects of subjective knowledge, objective knowledge, and usage experience on decision making: An exploratory investigation. *J. Consum. Psychol.* **1995**, *4*, 153–180. [CrossRef]

96. Faul, F.; Erdfelder, E.; Buchner, A.; Lang, A.-G. Statistical power analyses using G* Power 3.1: Tests for correlation and regression analyses. *Behav. Res. Methods* **2009**, *41*, 1149–1160. [CrossRef]

97. Vidaver-Cohen, D. Moral climate in business firms: A conceptual framework for analysis and change. *J. Bus. Ethics* **1998**, *17*, 1211–1226. [CrossRef]

98. Barclay, M.J.; Smith, C.W.; Watts, R.L. The determinants of corporate leverage and dividend policies. *J. Appl. Corp. Financ.* **1995**, *7*, 4–19. [CrossRef]

99. Ali, M.R.; Shafiq, M.; Andejany, M. Determinants of consumers' intentions towards the purchase of energy efficient appliances in Pakistan: An extended model of the theory of planned behavior. *Sustainability* **2021**, *13*, 565. [CrossRef]

100. Anderson, J.C.; Gerbing, D.W. Structural equation modeling in practice: A review and recommended two-step approach. *Psychol. Bull.* **1988**, *103*, 411. [CrossRef]

101. Fornell, C.; Larcker, D.F. Structural Equation Models with Unobservable Variables and Measurement Error: Algebra and Statistics. *J. Mark. Res.* **1981**, *18*, 382–388. [CrossRef]

102. Akter, S.; D'Ambra, J.; Ray, P. Trustworthiness in mHealth information services: An assessment of a hierarchical model with mediating and moderating effects using partial least squares (PLS). *J. Am. Soc. Inf. Sci. Technol.* **2011**, *62*, 100–116. [CrossRef]

103. Henseler, J.; Ringle, C.M.; Sarstedt, M. A new criterion for assessing discriminant validity in variance-based structural equation modeling. *J. Acad. Mark. Sci.* **2015**, *43*, 115–135. [CrossRef]

104. Kline, R.B. *Convergence of Structural Equation Modeling and Multilevel Modeling*; Sage Publishing: Thousand Oaks, CA, USA, 2011.

105. Kleinbaum, D.G.; Kupper, L.L.; Muller, K.E. Applied Regression Analysys and Other Multivariable Methods. In *Applied Regression Analysys and Other Multivariable Methods*; Duxbury Press: Pacific Grove, CA, USA, 1988; p. 718.

106. Cohen, J. *Statistical Power Analysis for the Behavioral Sciences*; Lawrence Erlbaum Associates: Hillsdale, NJ, USA, 1988; pp. 20–26.

107. Falk, R.F.; Miller, N.B. *A Primer for Soft Modeling*; University of Akron Press: Akron, OH, USA, 1992; ISBN 0962262846.

108. Harman, H.H. *Modern Factor Analysis*; University of Chicago Press: Chicago, IL, USA, 1976; ISBN 0226316521.

109. Holbert, R.L.; Stephenson, M.T. Structural Equation Modeling in the Communication Sciences, 1995–2000. *Hum. Commun. Res.* **2002**, *28*, 1995–2000. [CrossRef]

110. Bentler, P.M.; Bonett, D.G. Significance tests and goodness of fit in the analysis of covariance structures. *Psychol. Bull.* **1980**, *88*, 588–606. [CrossRef]

111. Jöreskog, K.G.; Sörbom, D. *LISREL 8: Structural Equation Modeling with the SIMPLIS Command Language*; Scientific Software International: Skokie, IL, USA, 1993.

112. McDonald, R.P.; Ho, M.-H.R. Principles and practice in reporting structural equation analyses. *Psychol. Methods* **2002**, *7*, 64. [CrossRef] [PubMed]

113. Bagozzi, R.P.; Yi, Y. On the evaluation of structural equation models. *J. Acad. Mark. Sci.* **1988**, *16*, 74–94. [CrossRef]

114. Browne, M.W.; Cudeck, R. Alternative ways of assessing model fit. In *Testing Structural Equation Models*; Bollen, K.A., Long, J.S., Eds.; Sage Publications: Thousand Oaks, CA, USA, 1993; pp. 136–162.

115. Si, H.; Shi, J.; Tang, D.; Wu, G.; Lan, J. Understanding intention and behavior toward sustainable usage of bike sharing by extending the theory of planned behavior. *Resour. Conserv. Recycl.* **2020**, *152*, 104513. [CrossRef]

116. Chen, A.; Peng, N. Recommending green hotels to travel agencies' customers. *Ann. Tour. Res.* **2014**, *48*, 284–289. [CrossRef]

117. Liu, P.; Teng, M.; Han, C. How does environmental knowledge translate into pro-environmental behaviors?: The mediating role of environmental attitudes and behavioral intentions. *Sci. Total Environ.* **2020**, *728*, 138126. [CrossRef]

118. Irianto, H. Consumers' attitude and intention towards organic food purchase: An extension of theory of planned behavior in gender perspective. *Int. J. Manag. Econ. Soc. Sci.* **2015**, *4*, 17–31.

119. Buabeng-Andoh, C. Predicting students' intention to adopt mobile learning: A combination of theory of reasoned action and technology acceptance model. *J. Res. Innov. Teach. Learn.* **2018**, *15*, 124–143. [CrossRef]

120. Zheng, G.-W.; Siddik, A.B.; Masukujjaman, M.; Alam, S.S.; Akter, A. Perceived Environmental Responsibilities and Green Buying Behavior: The Mediating Effect of Attitude. *Sustainability* **2021**, *13*, 35. [CrossRef]

MDPI
St. Alban-Anlage 66
4052 Basel
Switzerland
Tel. +41 61 683 77 34
Fax +41 61 302 89 18
www.mdpi.com

Applied Sciences Editorial Office
E-mail: applsci@mdpi.com
www.mdpi.com/journal/applsci